Introduction to Plant Reproduction

Introduction to Plant Reproduction

Edited by Seth Cohen

SYRAWOOD PUBLISHING HOUSE
New York

Published by Syrawood Publishing House,
750 Third Avenue, 9th Floor,
New York, NY 10017, USA
www.syrawoodpublishinghouse.com

Introduction to Plant Reproduction
Edited by Seth Cohen

International Standard Book Number: 978-1-68286-583-5 (Hardback)

Cataloging-in-Publication Data

Introduction to plant reproduction / edited by Seth Cohen.
p. cm.
Includes bibliographical references and index.
ISBN 978-1-68286-583-5
1. Plants--Reproduction. 2. Plant physiology. I. Cohen, Seth.
QK825 .I58 2018
575.6--dc23

Printed in China.

TABLE OF CONTENTS

PREFACE

As a part of plant science, plant reproduction is concerned with the study of production of new plants through asexual and sexual processes. It focuses on the crucial aspects of asexual reproduction, which include vegetative reproduction and apomixis and sexual reproduction, which include processes like meiosis and fertilization. This book presents the different concepts and methods related to the field of plant reproduction. Different approaches, evaluations and methodologies have been included in it. This textbook aims to serve as a resource guide for students and experts alike and contribute to the growth of the discipline.

Given below is the chapter wise description of the book:

Chapter 1- Plant reproduction is the production that occurs in plants. It can either be sexual or asexual reproduction. Flowering plant, stamen, stigma and vegetative reproduction are some of the significant and important topics related to plant reproduction. The following chapter unfolds its crucial aspects in a critical yet systematic manner.

Chapter 2- A seed usually consists of an embryo and a seed coat. Seeds store essential nutrients required for the growth of the plant. Seed dispersal, seedling and cotyledon are natural methods of plant propagation. The chapter on seed and germination process offers an insightful focus, keeping in mind the complex subject matter.

Chapter 3- The transfer of pollen from the male to the female part of the plant is known as pollination. There are two types of pollinations that occur, self-pollination and cross-pollination (xenogamy). A pollinator is an animal that aids in a plant's pollination. This chapter is an overview of the subject matter incorporating all the main aspects of pollination.

Chapter 4- Plant breeding is the act of changing certain traits of plants to generate preferred features. The techniques used in the process of plant breeding range from simply selecting desirable characteristics in plants to complex molecular techniques. Doubled haploidy, plant propagation and genetically modified crops are some of the topics related to plant breeding that have been discussed in the following section. The topics discussed in the chapter are of great importance to broaden the existing knowledge on plant reproduction.

Chapter 5- Micropropagation is a process of choosing the kind of plant material that needs to be propagated. The steps taken in this procedure are establishment, multiplication, pretransplant and transfer from culture. In order to completely understand micropropagation and plant hormone, it is necessary to understand the processes related to it. The following chapter elucidates the varied processes and mechanisms associated with this area of study.

At the end, I would like to thank all those who dedicated their time and efforts for the successful completion of this book. I also wish to convey my gratitude towards my friends and family who supported me at every step.

Editor

1

Understanding Plant Reproduction

Plant reproduction is the production that occurs in plants. It can either be sexual or asexual reproduction. Flowering plant, stamen, stigma and vegetative reproduction are some of the significant and important topics related to plant reproduction. The following chapter unfolds its crucial aspects in a critical yet systematic manner.

Plant Reproduction

Plant reproduction is the production of new individuals or offspring in plants, which can be accomplished by sexual or asexual reproduction. Sexual reproduction produces offspring by the fusion of gametes, resulting in offspring genetically different from the parent or parents. Asexual reproduction produces new individuals without the fusion of gametes, genetically identical to the parent plants and each other, except when mutations occur. In seed plants, the offspring can be packaged in a protective seed, which is used as an agent of dispersal.

Bryophyllum, a plant that reproduces asexually via new shoots from the leaves

Asexual Reproduction

Plants have two main types of asexual reproduction in which new plants are produced that are genetically identical clones of the parent individual. Vegetative reproduction involves a vegetative piece of the original plant (budding, tillering, etc.) and is distinguished from *apomixis*, which is a replacement for sexual reproduction, and in some cases involves seeds. Apomixis occurs in many plant species and also in some non-plant organisms. For apomixis and similar processes in non-plant organisms.

Natural vegetative reproduction is mostly a process found in herbaceous and woody perennial plants, and typically involves structural modifications of the stem or roots and in a few species leaves. Most plant species that employ vegetative reproduction do so as a means to perennialize the plants, allowing them to survive from one season to the next and often facilitating their expansion in size. A plant that persists in a location through vegetative reproduction of individuals constitutes a clonal colony; a single ramet, or apparent individual, of a clonal colony is genetically identical to all others in the same colony. The distance that a plant can move during vegetative reproduction is limited, though some plants can produce ramets from branching rhizomes or stolons that cover a wide area, often in only a few growing seasons. In a sense, this process is not one of reproduction but one of survival and expansion of biomass of the individual. When an individual organism increases in size via cell multiplication and remains intact, the process is called vegetative growth. However, in vegetative reproduction, the new plants that result are new individuals in almost every respect except genetic. A major disadvantage to vegetative reproduction, is the transmission of pathogens from parent to offspring; it is uncommon for pathogens to be transmitted from the plant to its seeds (in sexual reproduction or in apomixis), though there are occasions when it occurs.

Seeds generated by apomixis are a means of asexual reproduction, involving the formation and dispersal of seeds that do not originate from the fertilization of the embryos. Hawkweed (*Hieracium*), dandelion (*Taraxacum*), some Citrus (*Citrus*) and Kentucky blue grass (*Poa pratensis*) all use this form of asexual reproduction. Pseudogamy occurs in some plants that have apomictic seeds, where pollination is often needed to initiate embryo growth, though the pollen contributes no genetic material to the developing offspring. Other forms of apomixis occur in plants also, including the generation of a plantlet in replacement of a seed or the generation of bulbils instead of flowers, where new cloned individuals are produced.

Structures

A rhizome is a modified underground stem serving as an organ of vegetative reproduction; the growing tips of the rhizome can separate as new plants, e.g., polypody, iris, couch grass and nettles.

Prostrate aerial stems, called runners or stolons are important vegetative reproduction organs in some species, such as the strawberry, numerous grasses, and some ferns.

Adventitious buds form on roots near the ground surface, on damaged stems (as on the stumps of cut trees), or on old roots. These develop into above-ground stems and leaves. A form of budding called suckering is the reproduction or regeneration of a plant by shoots that arise from an existing root system. Species that characteristically produce suckers include Elm (*Ulmus*), Dandelion (*Taraxacum*), and many members of the Rose family such as *Rosa* and *Rubus*.

Plants like onion (*Allium cepa*), hyacinth (*Hyacinth*), narcissus (*Narcissus*) and tulips (*Tulipa*) reproduce by dividing their underground bulbs into more bulbs. Other plants like potatoes (*Solanum tuberosum*) and dahlia (*Dahlia*) reproduce by a similar method involving underground tubers. Gladioli and crocuses (*Crocus*) reproduce in a similar way with corms.

Usage

The most common form of plant reproduction utilized by people is seeds, but a number of asexual methods are utilized which are usually enhancements of natural processes, including: cutting, grafting, budding, layering, division, sectioning of rhizomes, roots, tubers, bulbs, stolons, tillers (suckers), etc., and artificial propagation by laboratory tissue cloning. Asexual methods are most often used to propagate cultivars with individual desirable characteristics that do not come true from seed. Fruit tree propagation is frequently performed by budding or grafting desirable cultivars (clones), onto rootstocks that are also clones, propagated by stooling.

In horticulture, a "cutting" is a branch that has been cut off from a mother plant below an internode and then rooted, often with the help of a rooting liquid or powder containing hormones. When a full root has formed and leaves begin to sprout anew, the clone is a self-sufficient plant, genetically identical to the mother plant. Examples include cuttings from the stems of blackberries (*Rubus occidentalis*), African violets (*Saintpaulia*), verbenas (*Verbena*) to produce new plants. A related use of cuttings is grafting, where a stem or bud is joined onto a different stem. Nurseries offer for sale trees with grafted stems that can produce four or more varieties of related fruits, including apples. The most common usage of grafting is the propagation of cultivars onto already rooted plants, sometimes the rootstock is used to dwarf the plants or protect them from root damaging pathogens.

Since vegetatively propagated plants are clones, they are important tools in plant research. When a clone is grown in various conditions, differences in growth can be ascribed to environmental effects instead of genetic differences.

Sexual Reproduction

Sexual reproduction involves two fundamental processes: meiosis, which rearranges the genes and reduces the number of chromosomes, and fertilization, which restores the chromosome to a complete diploid number. In between these two processes, different types of plants and algae vary, but many of them, including all land plants, undergo alternation of generations, with two different multicellular structures (phases), a gametophyte and a sporophyte. The evolutionary origin and adaptive significance of sexual reproduction are discussed in the pages "Evolution of sexual reproduction" and "Origin and function of meiosis."

The gametophyte is the multicellular structure (plant) that is haploid, containing a single set of chromosomes in each cell. The gametophyte produces male or female gametes (or both), by a process of cell division called mitosis. In vascular plants with separate gametophytes, female gametophytes are known as mega gametophytes (mega=large, they produce the large egg cells) and the male gametophytes are called micro gametophytes (micro=small, they produce the small sperm cells).

The fusion of male and female gametes (fertilization) produces a diploid zygote, which develops by mitotic cell divisions into a multicellular sporophyte.

The mature sporophyte produces spores by meiosis, sometimes referred to as "reduction division" because the chromosome pairs are separated once again to form single sets.

In mosses and liverworts the gametophyte is relatively large, and the sporophyte is a much smaller structure that is never separated from the gametophyte. In ferns, gymnosperms, and flowering plants (angiosperms), the gametophytes are relatively small and the sporophyte is much larger. In gymnosperms and flowering plants the mega gametophyte is contained within the ovule (that may develop into a seed) and the micro gametophyte is contained within a pollen grain.

History of Sexual Reproduction

Unlike animals, plants are immobile, and cannot seek out sexual partners for reproduction. In the evolution of early plants, abiotic means, including water and wind, transported sperm for reproduction. The first plants were aquatic, as described in the page "Evolutionary history of plants", and released sperm freely into the water to be carried with the currents. Primitive land plants like liverworts and mosses had motile sperm that swam in a thin film of water or were splashed in water droplets from the male reproduction organs onto the female organs. As taller and more complex plants evolved, modifications in the alternation of generations evolved; in the Paleozoic era progymnosperms reproduced by using spores dispersed on the wind. The seed plants including seed ferns, conifers and cordaites, which were all gymnosperms, evolved 350 million years ago; they had pollen grains that contained the male gametes for protection of the sperm during the process of transfer from the male to female parts. It is believed that insects fed on the pollen, and plants thus evolved to use insects to actively carry pollen from one plant to the next. Seed producing plants, which include the angiosperms and the gymnosperms, have heteromorphic alternation of generations with large sporophytes containing much reduced gametophytes. Angiosperms have distinctive reproductive organs called flowers, with carpels, and the female gametophyte is greatly reduced to a female embryo sac, with as few as eight cells. The male gametophyte consists of the pollen grains. The sperm of seed plants are non-motile, except for two older groups of plants, the Cycadophyta and the Ginkgophyta, which have flagellated sperm.

Flowering Plants

Flowering plants are the dominant plant form on land and they reproduce by sexual and asexual means. Often their most distinguishing feature is their reproductive organs, commonly called flowers. Sexual reproduction in flowering plants involves the production of male and female gametes, the transfer of the male gametes to the female ovules in a process called pollination. After pollination occurs, fertilization happens and the ovules grow into seeds within a fruit. After the seeds are ready for dispersal, the fruit ripens and by various means the seeds are freed from the fruit and after varying amounts of time and under specific conditions the seeds germinate and grow into the next generation.

The anther produces male gametophytes which are pollen grains, which attach to the stigma on top of a carpel, in which the female gametophytes (inside ovules) are located. After the pollen tube grows through the carpel's style, the sperm from the pollen grain migrate into the ovule to fertilize the egg cell and central cell within the female gametophyte in a process termed double fertilization. The resulting zygote develops into an embryo, while the triploid endosperm (one sperm cell plus a binucleate female cell) and female tissues of the ovule give rise to the surrounding tissues in the developing seed. The ovary, which produced the female gametophyte(s), then grows into a fruit, which surrounds the seed(s). Plants may either self-pollinate or cross-pollinate.

Pollination

In plants that use insects or other animals to move pollen from one flower to the next, plants have developed greatly modified flower parts to attract pollinators and to facilitate the movement of pollen from one flower to the insect and from the insect back to the next flower. Flowers of wind pollinated plants tend to lack petals and or sepals; typically large amounts of pollen are produced and pollination often occurs early in the growing season before leaves can interfere with the dispersal of the pollen. Many trees and all grasses and sedges are wind pollinated, as such they have no need for large fancy flowers.

An orchid flower

Plants have a number of different means to attract pollinators including colour, scent, heat, nectar glands, edible pollen and flower shape. Along with modifications involving the above structures two other conditions play a very important role in the sexual reproduction of flowering plants, the first is timing of flowering and the other is the size or number of flowers produced. Often plant species have a few large, very showy flowers while others produce many small flowers, often flowers are collected together into large inflorescences to maximize their visual effect, becoming more noticeable to passing pollinators. Flowers are attraction strategies and sexual expressions are functional strategies used to produce the next generation of plants, with pollinators and plants having co-evolved, often to some extraordinary degrees, very often rendering mutual benefit.

Flower heads showing disk and ray florets.

The largest family of flowering plants is the orchids (Orchidaceae), estimated by some specialists to include up to 35,000 species, which often have highly specialized flowers that attract particular insects for pollination. The stamens are modified to produce pollen in clusters called pollinia, which become attached to insects that crawl into the flower. The flower shapes may force insects to pass by the pollen, which is "glued" to the insect. Some orchids are even more highly specialized, with flower shapes that mimic the shape of insects to attract them to 'mate' with the flowers, a few even have scents that mimic insect pheromones.

Another large group of flowering plants is the Asteraceae or sunflower family with close to 22,000 species, which also have highly modified inflorescences that are flowers collected together in heads composed of a composite of individual flowers called florets. Heads with florets of one sex, when the flowers are pistillate or functionally staminate, or made up of all bisexual florets, are called homogamous and can include discoid and liguliflorous type heads. Some radiate heads may be homogamous too. Plants with heads that have florets of two or more sexual forms are called heterogamous and include radiate and disciform head forms, though some radiate heads may be heterogamous too.

Ferns

Ferns typically produce large diploid sporophytes with rhizomes, roots and leaves; and on fertile leaves called sporangium, spores are produced. The spores are released and germinate to produce short, thin gametophytes that are typically heart shaped, small and green in color. The gametophytes or thallus, produce both motile sperm in the antheridia and egg cells in separate archegonia. After rains or when dew deposits a film of water, the motile sperm are splashed away from the antheridia, which are normally produced on the top side of the thallus, and swim in the film of water to the antheridia where they fertilize the egg. To promote out crossing or cross fertilization the sperm are released before the eggs are receptive of the sperm, making it more likely that the sperm will fertilize the eggs of different thallus. A zygote is formed after fertilization, which grows into a new sporophytic plant. The condition of having separate sporophyte and gametophyte plants is call alternation of generations. Other plants with similar reproductive means include the *Psilotum*, *Lycopodium*, *Selaginella* and *Equisetum*.

Bryophytes

The bryophytes, which include liverworts, hornworts and mosses, reproduce both sexually and vegetatively. The gametophyte is the most commonly known phase of the plant. All are small plants found growing in moist locations and like ferns, have motile sperm with flagella and need water to facilitate sexual reproduction. These plants start as a haploid spore that grows into the dominate form, which is a multicellular haploid body with leaf-like structures that photosynthesize. Haploid gametes are produced in antherida and archegonia by mitosis. The sperm released from the antheridia respond to chemicals released by ripe archegonia and swim to them in a film of water and fertilize the egg cells, thus producing zygotes that are diploid. The zygote divides by mitotic division and grows into a sporophyte that is diploid. The multicellular diploid sporophyte produces structures called spore capsules. The spore capsules produce spores by meiosis, and when ripe, the capsules burst open and the spores are released. Bryophytes show considerable variation in their breeding structures and the above is a basic outline. In some species each gametophyte is one sex while other species produce both antheridia and archegonia on the same gametophyte which is thus hermaphrodite.

Sexual Morphology

Many plants have evolved complex sexual reproductive systems, which is expressed in different combinations of their reproductive organs. Some species have separate male and female plants, and some have separate male and female flowers on the same plant, but the majority of plants have both male and female parts in the same flower. Some plants change their morphological expression depending on a number of factors like age, time of day, or because of environmental conditions. Plant sexual morphology also varies within different populations of some species.

Flowering Plant

The flowering plants (angiosperms), also known as Angiospermae or Magnoliophyta, are the most diverse group of land plants, with 416 families, approx. 13,164 known genera and a total of c. 295,383 known species. Like gymnosperms, angiosperms are seed-producing plants; they are distinguished from gymnosperms by characteristics including flowers, endosperm within the seeds, and the production of fruits that contain the seeds. Etymologically, angiosperm means a plant that produces seeds within an enclosure, in other words, a fruiting plant. The term "angiosperm" comes from the Greek composite word (*angeion*, "case" or "casing", and *sperma*, "seed") meaning "enclosed seeds", after the enclosed condition of the seeds.

The ancestors of flowering plants diverged from gymnosperms in the Triassic Period, during the range 245 to 202 million years ago (mya), and the first flowering plants are known from 160 mya. They diversified extensively during the Lower Cretaceous, became widespread by 120 mya, and replaced conifers as the dominant trees from 100 to 60 mya.

Description

Angiosperm Derived Characteristics

Bud of a pink rose

Angiosperms differ from other seed plants in several ways, described in the table. These distinguishing characteristics taken together have made the angiosperms the most diverse and numerous land plants and the most commercially important group to humans.

Distinctive features of Angiosperms	
Feature	**Description**
Flowering organs	Flowers, the reproductive organs of flowering plants, are the most remarkable feature distinguishing them from the other seed plants. Flowers provided angiosperms with the means to have a more species-specific breeding system, and hence a way to evolve more readily into different species without the risk of crossing back with related species. Faster speciation enabled the Angiosperms to adapt to a wider range of ecological niches. This has allowed flowering plants to largely dominate terrestrial ecosystems.
Stamens with two pairs of pollen sacs	Stamens are much lighter than the corresponding organs of gymnosperms and have contributed to the diversification of angiosperms through time with adaptations to specialized pollination syndromes, such as particular pollinators. Stamens have also become modified through time to prevent self-fertilization, which has permitted further diversification, allowing angiosperms eventually to fill more niches.
Reduced male parts, three cells	The male gametophyte in angiosperms is significantly reduced in size compared to those of gymnosperm seed plants. The smaller size of the pollen reduces the amount of time between pollination — the pollen grain reaching the female plant — and fertilization. In gymnosperms, fertilization can occur up to a year after pollination, whereas in angiosperms, fertilization begins very soon after pollination. The shorter amount of time between pollination and fertilization allows angiosperms to produce seeds earlier after pollination than gymnosperms, providing angiosperms a distinct evolutionary advantage.
Closed carpel enclosing the ovules (carpel or carpels and accessory parts may become the fruit)	The closed carpel of angiosperms also allows adaptations to specialized pollination syndromes and controls. This helps to prevent self-fertilization, thereby maintaining increased diversity. Once the ovary is fertilized, the carpel and some surrounding tissues develop into a fruit. This fruit often serves as an attractant to seed-dispersing animals. The resulting cooperative relationship presents another advantage to angiosperms in the process of dispersal.
Reduced female gametophyte, seven cells with eight nuclei	The reduced female gametophyte, like the reduced male gametophyte, may be an adaptation allowing for more rapid seed set, eventually leading to such flowering plant adaptations as annual herbaceous life-cycles, allowing the flowering plants to fill even more niches.
Endosperm	In general, endosperm formation begins after fertilization and before the first division of the zygote. Endosperm is a highly nutritive tissue that can provide food for the developing embryo, the cotyledons, and sometimes the seedling when it first appears.

Vascular Anatomy

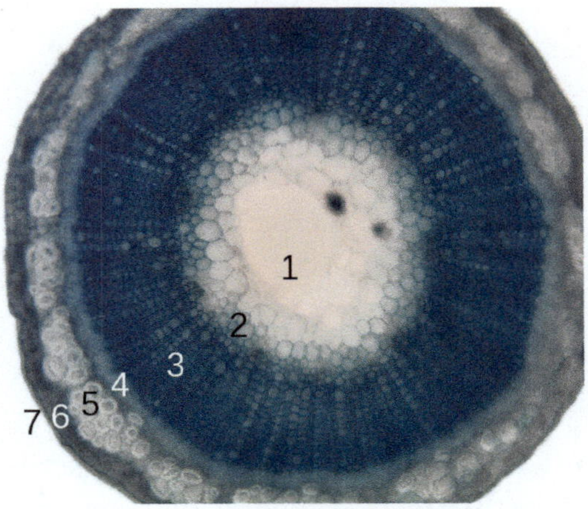

Cross-section of a stem of the angiosperm flax:
1. Pith, 2. Protoxylem, 3. Xylem I, 4. Phloem I, 5. Sclerenchyma (bast fibre), 6. Cortex, 7. Epidermis

The amount and complexity of tissue-formation in flowering plants exceeds that of gymnosperms. The vascular bundles of the stem are arranged such that the xylem and phloem form concentric rings.

In the dicotyledons, the bundles in the very young stem are arranged in an open ring, separating a central pith from an outer cortex. In each bundle, separating the xylem and phloem, is a layer of meristem or active formative tissue known as cambium. By the formation of a layer of cambium between the bundles (interfascicular cambium), a complete ring is formed, and a regular periodical increase in thickness results from the development of xylem on the inside and phloem on the outside. The soft phloem becomes crushed, but the hard wood persists and forms the bulk of the stem and branches of the woody perennial. Owing to differences in the character of the elements produced at the beginning and end of the season, the wood is marked out in transverse section into concentric rings, one for each season of growth, called annual rings.

Among the monocotyledons, the bundles are more numerous in the young stem and are scattered through the ground tissue. They contain no cambium and once formed the stem increases in diameter only in exceptional cases.

Reproductive Anatomy

A collection of flowers forming an inflorescence

The characteristic feature of angiosperms is the flower. Flowers show remarkable variation in form and elaboration, and provide the most trustworthy external characteristics for establishing relationships among angiosperm species. The function of the flower is to ensure fertilization of the ovule and development of fruit containing seeds. The floral apparatus may arise terminally on a shoot or from the axil of a leaf (where the petiole attaches to the stem). Occasionally, as in violets, a flower arises singly in the axil of an ordinary foliage-leaf. More typically, the flower-bearing portion of the plant is sharply distinguished from the foliage-bearing or vegetative portion, and forms a more or less elaborate branch-system called an inflorescence.

There are two kinds of reproductive cells produced by flowers. Microspores, which will divide to become pollen grains, are the "male" cells and are borne in the stamens (or microsporophylls). The "female" cells called megaspores, which will divide to become the egg cell (megagametogenesis), are contained in the ovule and enclosed in the carpel (or megasporophyll).

The flower may consist only of these parts, as in willow, where each flower comprises only a few stamens or two carpels. Usually, other structures are present and serve to protect the sporophylls and to form an envelope attractive to pollinators. The individual members of these surrounding structures are known as sepals and petals (or tepals in flowers such as *Magnolia* where sepals and petals are not distinguishable from each other). The outer series (calyx of sepals) is usually green and leaf-like, and functions to protect the rest of the flower, especially the bud. The inner series (corolla of petals) is, in general, white or brightly colored, and is more delicate in structure. It functions to attract insect or bird pollinators. Attraction is effected by color, scent, and nectar, which may be secreted in some part of the flower. The characteristics that attract pollinators account for the popularity of flowers and flowering plants among humans.

While the majority of flowers are perfect or hermaphrodite (having both pollen and ovule producing parts in the same flower structure), flowering plants have developed numerous morphological and physiological mechanisms to reduce or prevent self-fertilization. Heteromorphic flowers have short carpels and long stamens, or vice versa, so animal pollinators cannot easily transfer pollen to the pistil (receptive part of the carpel). Homomorphic flowers may employ a biochemical (physiological) mechanism called self-incompatibility to discriminate between self and non-self pollen grains. In other species, the male and female parts are morphologically separated, developing on different flowers.

Taxonomy

History of Classification

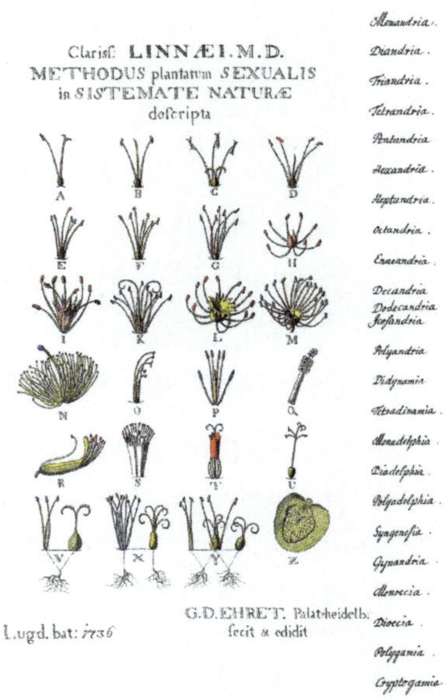

From 1736, an illustration of Linnaean classification

The botanical term "Angiosperm", from the Ancient Greek, *angeíon* (bottle, vessel) and (seed), was coined in the form Angiospermae by Paul Hermann in 1690, as the name of one of his primary divisions of the plant kingdom. This included flowering plants possessing seeds enclosed in capsules, distinguished from his Gymnospermae, or flowering plants with achenial or schizo-carpic fruits, the whole fruit or each of its pieces being here regarded as a seed and naked. The term and its antonym were maintained by Carl Linnaeus with the same sense, but with restricted application, in the names of the orders of his class Didynamia. Its use with any approach to its modern scope became possible only after 1827, when Robert Brown established the existence of truly naked ovules in the Cycadeae and Coniferae, and applied to them the name Gymnosperms. From that time onward, as long as these Gymnosperms were, as was usual, reckoned as dicotyledonous flowering plants, the term Angiosperm was used antithetically by botanical writers, with varying scope, as a group-name for other dicotyledonous plants.

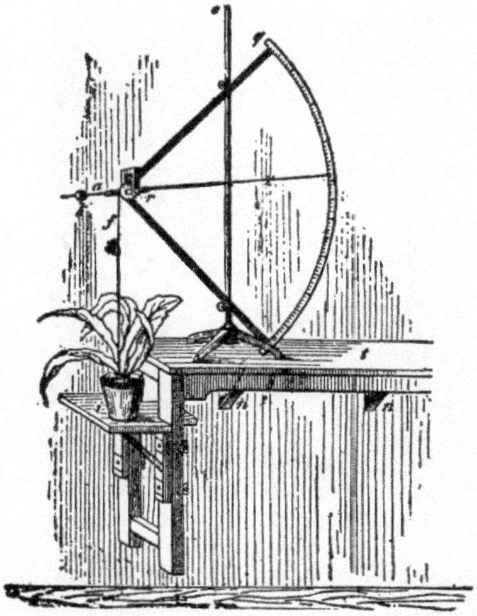

An auxanometer, a device for measuring increase or rate of growth in plants

In 1851, Hofmeister discovered the changes occurring in the embryo-sac of flowering plants, and determined the correct relationships of these to the Cryptogamia. This fixed the position of Gymnosperms as a class distinct from Dicotyledons, and the term Angiosperm then gradually came to be accepted as the suitable designation for the whole of the flowering plants other than Gymnosperms, including the classes of Dicotyledons and Monocotyledons. This is the sense in which the term is used today.

In most taxonomies, the flowering plants are treated as a coherent group. The most popular descriptive name has been Angiospermae (Angiosperms), with Anthophyta ("flowering plants") a second choice. These names are not linked to any rank. The Wettstein system and the Engler system use the name Angiospermae, at the assigned rank of subdivision. The Reveal system treated flowering plants as subdivision Magnoliophytina (Frohne & U. Jensen ex Reveal, Phytologia 79: 70 1996), but later split it to Magnoliopsida, Liliopsida, and Rosopsida. The Takhtajan system and Cronquist system treat this group at the rank of division, leading to the name Magnoliophyta (from the family name Magnoliaceae). The Dahlgren system and Thorne system (1992) treat this group at the rank of class, leading to the name Magnoliopsida. The APG system of 1998, and the later 2003 and 2009 revisions, treat the flowering plants as a clade called angiosperms without a formal botanical name. However, a formal classification was published alongside the 2009 revision in which the flowering plants form the Subclass Magnoliidae.

The internal classification of this group has undergone considerable revision. The Cronquist system, proposed by Arthur Cronquist in 1968 and published in its full form in 1981, is still widely used but is no longer believed to accurately reflect phylogeny. A consensus about how the flowering plants should be arranged has recently begun to emerge through the work of the Angiosperm Phylogeny Group (APG), which published an influential reclassification of the angiosperms in 1998. Updates incorporating more recent research were published as APG II in 2003 and as APG III in 2009.

Monocot (left) and dicot seedlings

Traditionally, the flowering plants are divided into two groups,

- Dicotyledoneae or Magnoliopsida
- Monocotyledoneae or Liliopsida

which in the Cronquist system are called Magnoliopsida (at the rank of class, formed from the family name Magnoliaceae) and Liliopsida (at the rank of class, formed from the family name Liliaceae). Other descriptive names allowed by Article 16 of the ICBN include Dicotyledones or Dicotyledoneae, and Monocotyledones or Monocotyledoneae, which have a long history of use. In English a member of either group may be called a dicotyledon (plural dicotyledons) and monocotyledon (plural monocotyledons), or abbreviated, as dicot (plural dicots) and monocot (plural monocots). These names derive from the observation that the dicots most often have two cotyledons, or embryonic leaves, within each seed. The monocots usually have only one, but the rule is not absolute either way. From a broad diagnostic point of view, the number of cotyledons is neither a particularly handy nor a reliable character.

Recent studies, as by the APG, show that the monocots form a monophyletic group (clade) but that the dicots do not (they are paraphyletic). Nevertheless, the majority of dicot species do form a monophyletic group, called the eudicots or tricolpates. Of the remaining dicot species, most belong to a third major clade known as the magnoliids, containing about 9,000 species. The rest include a paraphyletic grouping of early branching taxa known collectively as the basal angiosperms, plus the families Ceratophyllaceae and Chloranthaceae.

Modern Classification

There are eight groups of living angiosperms:

- Basal angiosperms (ANA: *Amborella*, Nymphaeales, Austrobaileyales)
    - *Amborella*, a single species of shrub from New Caledonia;
    - Nymphaeales, about 80 species, water lilies and Hydatellaceae;
    - Austrobaileyales, about 100 species of woody plants from various parts of the world
- Core angiosperms (Mesangiospermae)
    - Chloranthales, several dozen species of aromatic plants with toothed leaves;
    - Magnoliids, about 9,000 species, characterized by trimerous flowers, pollen with one pore, and usually branching-veined leaves—for example magnolias, bay laurel, and black pepper;
    - Monocots, about 70,000 species, characterized by trimerous flowers, a single cotyledon, pollen with one pore, and usually parallel-veined leaves—for example grasses, orchids, and palms;
    - *Ceratophyllum*, about 6 species of aquatic plants, perhaps most familiar as aquarium plants;
    - Eudicots, about 175,000 species, characterized by 4- or 5-merous flowers, pollen with three pores, and usually branching-veined leaves—for example sunflowers, petunia, buttercup, apples, and oaks.

Evolution

Flowers of *Malus sylvestris* (crab apple)

Fossilized spores suggest that higher plants (embryophytes) have lived on land for at least 475 million years. Early land plants reproduced sexually with flagellated, swimming sperm, like the green algae from which they evolved. An adaptation to terrestrialization was the development of upright meiosporangia for dispersal by spores to new habitats. This feature is lacking in the descendants of their nearest algal relatives, the Charophycean green algae. A later terrestrial adaptation took place with retention of the delicate, avascular sexual stage, the gametophyte, within the tissues of the vascular sporophyte. This occurred by spore germination within sporangia rather than spore release, as in non-seed plants. A current example of how this might have happened can be seen in the precocious spore germination in *Selaginella*, the spike-moss. The result for the ancestors of angiosperms was enclosing them in a case, the seed. The first seed bearing plants, like the ginkgo, and conifers (such as pines and firs), did not produce flowers. The pollen grains (male gametophytes) of *Ginkgo* and cycads produce a pair of flagellated, mobile sperm cells that "swim" down the developing pollen tube to the female and her eggs.

Flowers and leaves of *Oxalis pes-caprae* (Bermuda buttercup)

The apparently sudden appearance of nearly modern flowers in the fossil record initially posed such a problem for the theory of evolution that Charles Darwin called it an "*abominable mystery*". However, the fossil record has considerably grown since the time of Darwin, and recently discovered angiosperm fossils such as *Archaefructus*, along with further discoveries of fossil gymnosperms, suggest how angiosperm characteristics may have been acquired in a series of steps. Several groups of extinct gymnosperms, in particular seed ferns, have been proposed as the ancestors of flowering plants, but there is no continuous fossil evidence showing exactly how flowers evolved. Some older fossils, such as the upper Triassic *Sanmiguelia*, have been suggested. Based on current evidence, some propose that the ancestors of the angiosperms diverged from an unknown group of gymnosperms in the Triassic period (245–202 million years ago). Fossil angiosperm-like pollen from the Middle Triassic (247.2–242.0 Ma) suggests an older date for their origin. A close relationship between angiosperms and gnetophytes, proposed on the basis of morphological evidence, has more recently been disputed on the basis of molecular evidence that suggest gnetophytes are instead more closely related to other gymnosperms.

The evolution of seed plants and later angiosperms appears to be the result of two distinct rounds of whole genome duplication events. These occurred at 319 million years ago and 192 million years ago. Another possible whole genome duplication event at 160 million years ago perhaps created the ancestral line that led to all modern flowering plants. That event was studied by sequencing the genome of an ancient flowering plant, *Amborella trichopoda,* and directly addresses Darwin's "*abominable mystery.*"

The earliest known macrofossil confidently identified as an angiosperm, *Archaefructus liaoningensis,* is dated to about 125 million years BP (the Cretaceous period), whereas pollen considered to be of angiosperm origin takes the fossil record back to about 130 million years BP. However, one study has suggested that the early-middle Jurassic plant *Schmeissneria,* traditionally considered a type of ginkgo, may be the earliest known angiosperm, or at least a close relative. In addition, circumstantial chemical evidence has been found for the existence of angiosperms as early as 250 million years ago. Oleanane, a secondary metabolite produced by many flowering plants, has been found in Permian deposits of that age together with fossils of gigantopterids. Gigantopterids are a group of extinct seed plants that share many morphological traits with flowering plants, although they are not known to have been flowering plants themselves.

In 2013 flowers encased in amber were found and dated 100 million years before present. The amber had frozen the act of sexual reproduction in the process of taking place. Microscopic images showed tubes growing out of pollen and penetrating the flower's stigma. The pollen was sticky, suggesting it was carried by insects.

Recent DNA analysis based on molecular systematics showed that *Amborella trichopoda,* found on the Pacific island of New Caledonia, belongs to a sister group of the other flowering plants, and morphological studies suggest that it has features that may have been characteristic of the earliest flowering plants.

The orders Amborellales, Nymphaeales, and Austrobaileyales diverged as separate lineages from the remaining angiosperm clade at a very early stage in flowering plant evolution.

The great angiosperm radiation, when a great diversity of angiosperms appears in the fossil record, occurred in the mid-Cretaceous (approximately 100 million years ago). However, a study in 2007 estimated that the division of the five most recent (the genus *Ceratophyllum,* the family Chloranthaceae, the eudicots, the magnoliids, and the monocots) of the eight main groups occurred around 140 million years ago. By the late Cretaceous, angiosperms appear to have dominated environments formerly occupied by ferns and cycadophytes, but large canopy-forming trees replaced conifers as the dominant trees only close to the end of the Cretaceous 66 million years ago or even later, at the beginning of the Tertiary. The radiation of herbaceous angiosperms occurred much later. Yet, many fossil plants recognizable as belonging to modern families (including beech, oak, maple, and magnolia) had already appeared by the late Cretaceous.

It is generally assumed that the function of flowers, from the start, was to involve mobile animals in their reproduction processes. That is, pollen can be scattered even if the flower is not brightly colored or oddly shaped in a way that attracts animals; however, by expending the energy required to create such traits, angiosperms can enlist the aid of animals and, thus, reproduce more efficiently.

Two bees on a flower head of Creeping Thistle, *Cirsium arvense*

Island genetics provides one proposed explanation for the sudden, fully developed appearance of flowering plants. Island genetics is believed to be a common source of speciation in general, especially when it comes to radical adaptations that seem to have required inferior transitional forms. Flowering plants may have evolved in an isolated setting like an island or island chain, where the plants bearing them were able to develop a highly specialized relationship with some specific animal (a wasp, for example). Such a relationship, with a hypothetical wasp carrying pollen from one plant to another much the way fig wasps do today, could result in the development of a high degree of specialization in both the plant(s) and their partners. Note that the wasp example is not incidental; bees, which, it is postulated, evolved specifically due to mutualistic plant relationships, are descended from wasps.

Animals are also involved in the distribution of seeds. Fruit, which is formed by the enlargement of flower parts, is frequently a seed-dispersal tool that attracts animals to eat or otherwise disturb it, incidentally scattering the seeds it contains. Although many such mutualistic relationships remain too fragile to survive competition and to spread widely, flowering proved to be an unusually effective means of reproduction, spreading (whatever its origin) to become the dominant form of land plant life.

Flower ontogeny uses a combination of genes normally responsible for forming new shoots. The most primitive flowers probably had a variable number of flower parts, often separate from (but in contact with) each other. The flowers tended to grow in a spiral pattern, to be bisexual (in plants, this means both male and female parts on the same flower), and to be dominated by the ovary (female part). As flowers evolved, some variations developed parts fused together, with a much more specific number and design, and with either specific sexes per flower or plant or at least "ovary-inferior".

Flower evolution continues to the present day; modern flowers have been so profoundly influenced by humans that some of them cannot be pollinated in nature. Many modern domesticated flower species were formerly simple weeds, which sprouted only when the ground was disturbed. Some of them tended to grow with human crops, perhaps already having symbiotic companion plant relationships with them, and the prettiest did not get plucked because of their beauty, developing a dependence upon and special adaptation to human affection.

A few paleontologists have also proposed that flowering plants, or angiosperms, might have evolved due to interactions with dinosaurs. One of the idea's strongest proponents is Robert T.

Bakker. He proposes that herbivorous dinosaurs, with their eating habits, provided a selective pressure on plants, for which adaptations either succeeded in deterring or coping with predation by herbivores.

Flowering Plant Diversity

A poster of twelve different species of flowers of the *Asteraceae* family

Lupinus pilosus

The number of species of flowering plants is estimated to be in the range of 250,000 to 400,000. This compares to around 12,000 species of moss or 11,000 species of pteridophytes, showing that the flowering plants are much more diverse. The number of families in APG (1998) was 462. In APG II (2003) it is not settled; at maximum it is 457, but within this number there are 55 optional segregates, so that the minimum number of families in this system is 402. In APG III (2009) there are 415 families.

The diversity of flowering plants is not evenly distributed. Nearly all species belong to the eudicot (75%), monocot (23%), and magnoliid (2%) clades. The remaining 5 clades contain a little over

250 species in total; i.e. less than 0.1% of flowering plant diversity, divided among 9 families. The 42 most-diverse of 443 families of flowering plants by species, in their APG circumscriptions, are

1. Asteraceae or Compositae (daisy family): 22,750 species;
2. Orchidaceae (orchid family): 21,950;
3. Fabaceae or Leguminosae (bean family): 19,400;
4. Rubiaceae (madder family): 13,150;
5. Poaceae or Gramineae (grass family): 10,035;
6. Lamiaceae or Labiatae (mint family): 7,175;
7. Euphorbiaceae (spurge family): 5,735;
8. Melastomataceae or Melastomaceae (melastome family): 5,005;
9. Myrtaceae (myrtle family): 4,625;
10. Apocynaceae (dogbane family): 4,555;
11. Cyperaceae (sedge family): 4,350;
12. Malvaceae (mallow family): 4,225;
13. Araceae (arum family): 4,025;
14. Ericaceae (heath family): 3,995;
15. Gesneriaceae (gesneriad family): 3,870;
16. Apiaceae or Umbelliferae (parsley family): 3,780;
17. Brassicaceae or Cruciferae (cabbage family): 3,710:
18. Piperaceae (pepper family): 3,600;
19. Acanthaceae (acanthus family): 3,500;
20. Rosaceae (rose family): 2,830;
21. Boraginaceae (borage family): 2,740;
22. Urticaceae (nettle family): 2,625;
23. Ranunculaceae (buttercup family): 2,525;
24. Lauraceae (laurel family): 2,500;
25. Solanaceae (nightshade family): 2,460;
26. Campanulaceae (bellflower family): 2,380;
27. Arecaceae (palm family): 2,361;

28. Annonaceae (custard apple family): 2,220;
29. Caryophyllaceae (pink family): 2,200;
30. Orobanchaceae (broomrape family): 2,060;
31. Amaranthaceae (amaranth family): 2,050;
32. Iridaceae (iris family): 2,025;
33. Aizoaceae or Ficoidaceae (ice plant family): 2,020;
34. Rutaceae (rue family): 1,815;
35. Phyllanthaceae (phyllanthus family): 1,745;
36. Scrophulariaceae (figwort family): 1,700;
37. Gentianaceae (gentian family): 1,650;
38. Convolvulaceae (bindweed family): 1,600;
39. Proteaceae (protea family): 1,600;
40. Sapindaceae (soapberry family): 1,580;
41. Cactaceae (cactus family): 1,500;
42. Araliaceae (*Aralia* or ivy family): 1,450.

Of these, the Orchidaceae, Poaceae, Cyperaceae, Arecaceae, and Iridaceae are monocot families; Piperaceae, Lauraceae, and Annonaceae are magnoliid dicots; the rest of the families are eudicots.

Ecology

Fertilization and Embryogenesis

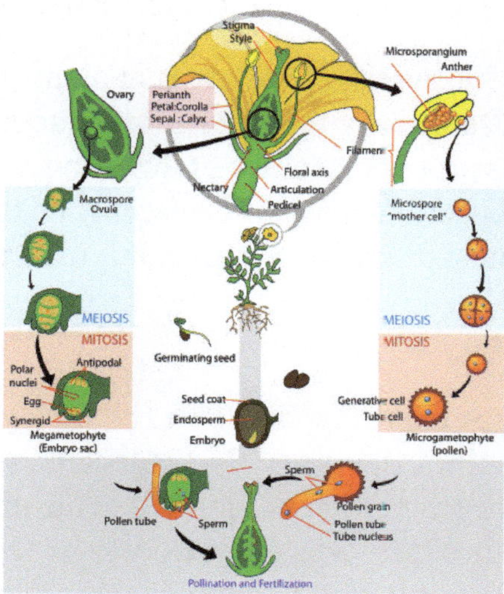

Angiosperm life cycle

Double fertilization refers to a process in which two sperm cells fertilize cells in the ovary. This process begins when a pollen grain adheres to the stigma of the pistil (female reproductive structure), germinates, and grows a long pollen tube. While this pollen tube is growing, a haploid generative cell travels down the tube behind the tube nucleus. The generative cell divides by mitosis to produce two haploid (n) sperm cells. As the pollen tube grows, it makes its way from the stigma, down the style and into the ovary. Here the pollen tube reaches the micropyle of the ovule and digests its way into one of the synergids, releasing its contents (which include the sperm cells). The synergid that the cells were released into degenerates and one sperm makes its way to fertilize the egg cell, producing a diploid ($2n$) zygote. The second sperm cell fuses with both central cell nuclei, producing a triploid ($3n$) cell. As the zygote develops into an embryo, the triploid cell develops into the endosperm, which serves as the embryo's food supply. The ovary will now develop into a fruit and the ovule will develop into a seed.

Fruit and Seed

The fruit of the *Aesculus* or Horse Chestnut tree

As the development of embryo and endosperm proceeds within the embryo sac, the sac wall enlarges and combines with the nucellus (which is likewise enlarging) and the integument to form the *seed coat*. The ovary wall develops to form the fruit or pericarp, whose form is closely associated with the manner of distribution of the seed.

Frequently, the influence of fertilization is felt beyond the ovary, and other parts of the flower take part in the formation of the fruit, e.g., the floral receptacle in the apple, strawberry, and others.

The character of the seed coat bears a definite relation to that of the fruit. They protect the embryo and aid in dissemination; they may also directly promote germination. Among plants with indehiscent fruits, in general, the fruit provides protection for the embryo and secures dissemination. In this case, the seed coat is only slightly developed. If the fruit is dehiscent and the seed is exposed, in general, the seed-coat is well developed, and must discharge the functions otherwise executed by the fruit.

Meiosis

Flowering plants generate gametes using a specialized cell division called meiosis. Meiosis takes place in the ovule (a structure within the ovary that is located within the pistil at the center of the flower). A diploid cell (megaspore mother cell) in the ovule undergoes meiosis (involving two successive cell divisions) to produce four cells (megaspores or female gametes) with haploid nuclei. One of these four cells (megaspore) then undergoes three successive mitotic divisions to produce an immature embryo sac (megagametocyte) with eight haploid nuclei. Next, these nuclei are segregated into separate cells by cytokinesis to producing 3 antipodal cells, 2 synergid cells and an egg cell. Two polar nuclei are left in the central cell of the embryo sac.

Pollen is also produced by meiosis in the male anther (microsporangium). During meiosis, a diploid microspore mother cell undergoes two successive meiotic divisions to produce 4 haploid cells (microspores or male gametes). Each of these microspores, after further mitoses, becomes a pollen grain (microgametophyte) containing two haploid generative (sperm) cells and a tube nucleus. When a pollen grain makes contact with the female stigma, the pollen grain forms a pollen tube that grows down the style into the ovary. In the act of fertilization, a male sperm nucleus fuses with the female egg nucleus to form a diploid zygote that can then develop into an embryo within the newly forming seed. Upon germination of the seed, a new plant can grow and mature.

The adaptive function of meiosis is currently a matter of debate. A key event during meiosis in a diploid cell is the pairing of homologous chromosomes and homologous recombination (the exchange of genetic information) between homologous chromosomes. This process promotes the production of increased genetic diversity among progeny and the recombinational repair of damages in the DNA to be passed on to progeny. To explain the adaptive function of meiosis in flowering plants, some authors emphasize diversity and others emphasize DNA repair.

Apomixis

Apomixis (reproduction via asexually formed seeds) is found naturally in about 2.2% of angiosperm genera One type of apomixis, gametophytic apomixis found in a dandelion species involves formation of an unreduced embryo sac due to incomplete meiosis (apomeiosis) and development of an embryo from the unreduced egg inside the embryo sac, without fertilization (parthenogenesis).

Uses

Agriculture is almost entirely dependent on angiosperms, which provide virtually all plant-based food, and also provide a significant amount of livestock feed. Of all the families of plants, the Poaceae, or grass family (grains), is by far the most important, providing the bulk of all feedstocks (rice, corn — maize, wheat, barley, rye, oats, pearl millet, sugar cane, sorghum). The Fabaceae, or legume family, comes in second place. Also of high importance are the Solanaceae, or nightshade family (potatoes, tomatoes, and peppers, among others), the Cucurbitaceae, or gourd family (also including pumpkins and melons), the Brassicaceae, or mustard plant family (including rapeseed and the innumerable varieties of the cabbage species *Brassica oleracea*), and the Apiaceae, or parsley family. Many of our fruits come from the Rutaceae, or rue family (including oranges, lemons, grapefruits, etc.), and the Rosaceae, or rose family (including apples, pears, cherries, apricots, plums, etc.).

In some parts of the world, certain single species assume paramount importance because of their variety of uses, for example the coconut (*Cocos nucifera*) on Pacific atolls, and the olive (*Olea europaea*) in the Mediterranean region.

Flowering plants also provide economic resources in the form of wood, paper, fiber (cotton, flax, and hemp, among others), medicines (digitalis, camphor), decorative and landscaping plants, and many other uses. The main area in which they are surpassed by other plants — namely, coniferous trees (Pinales), which are non-flowering (gymnosperms) — is timber and paper production.

Double Fertilization

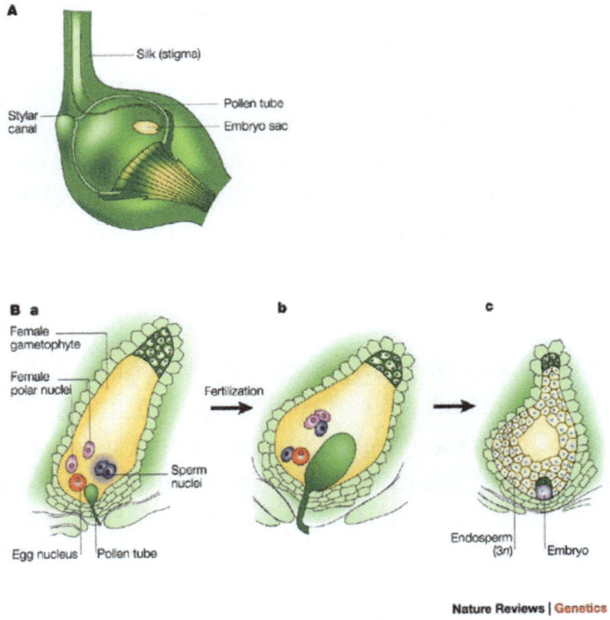

Double fertilization

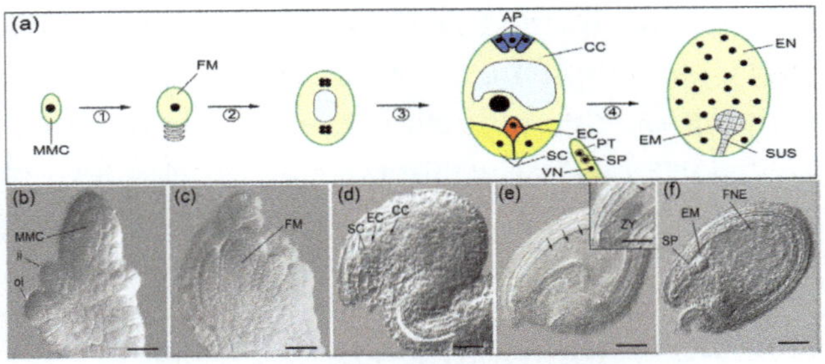

Double fertilization in *Arabidopsis*

Double fertilization is a complex fertilization mechanism of flowering plants (angiosperms). This process involves the joining of a female gametophyte (megagametophyte, also called the embryo sac) with two male gametes (sperm). It begins when a pollen grain adheres to the stigma of the carpel, the female reproductive structure of a flower. The pollen grain then takes in moisture and begins to germinate, forming a pollen tube that extends down toward the ovary through the style. The tip of the pollen tube then enters the ovary and penetrates through the micropyle opening in the ovule. The pollen tube proceeds to release the two sperm in the megagametophyte.

One sperm fertilizes the egg cell and the other sperm combines with the two polar nuclei of the large central cell of the megagametophyte. The haploid sperm and haploid egg combine to form a diploid zygote, while the other sperm and the two haploid polar nuclei of the large central cell of the megagametophyte form a triploid nucleus (triple fusion). Some plants may form polyploid nuclei. The large cell of the gametophyte will then develop into the endosperm, a nutrient-rich tissue which provides nourishment to the developing embryo. The ovary, surrounding the ovules, develops into the fruit, which protects the seeds and may function to disperse them.

The two central cell maternal nuclei (polar nuclei) that contribute to the endosperm, arise by mitosis from the same single meiotic product that gave rise to the egg. The maternal contribution to the genetic constitution of the triploid endosperm is double that of the embryo.

In a study conducted in 2008 of the plant *Arabidopsis thaliana*, the migration of male nuclei inside the female gamete, in fusion with the female nuclei, has been documented for the first time using in vivo imaging. Some of the genes involved in the migration and fusion process have also been determined.

Evidence of double fertilization in Gnetales, which are non-flowering seed plants, has been reported.

Brief History

Double fertilization was discovered more than a century ago by Sergei Nawaschin in Kiev, Russian Empire, and Léon Guignard in France. Each made the discovery independently of the other. *Lilium martagon* and *Fritillaria tenella* were used in the first observations of double fertilization, which were made using the classical light microscope. Due to the limitations of the light microscope, there were many unanswered questions regarding the process of double fertilization. However, with the development of the electron microscope, many of the questions were answered. Most notably, the observations made by the group of W. Jensen showed that the male gametes did not have any cell walls and that the plasma membrane of the gametes is close to the plasma membrane of the cell that surrounds them inside the pollen grain.

In Vitro Double Fertilization

In vitro double fertilization is often used to study the molecular interactions as well as other aspects of gamete fusion in flowering plants. One of the major obstacles in developing an in vitro double fertilization between male and female gametes is the confinement of the sperm in the pollen tube and the egg in the embryo sac. A controlled fusion of the egg and sperm has already been achieved with poppy plants. Pollen germination, pollen tube entry, and double fertilization processes have all been observed to proceed normally. In fact, this technique has already been used to obtain seeds in various flowering plants and was named "test-tube fertilization".

Related Structures and Functions

Megagametophyte

The female gametophyte, the megagametophyte, that participates in double fertilization in angiosperms is sometimes called the embryo sac. This develops within an ovule, enclosed by the ovary at the base of a carpel. Surrounding the megagametophyte are (one or) two integuments, which form an opening called the micropyle. The megagametophyte, which is usually haploid, originates from the (usually diploid) megaspore mother cell, also called the megasporocyte. The next sequence of events varies, depending on the particular species, but in most species, the following events occur. The megasporocyte undergoes a meiotic cell division, producing four haploid megaspores. Only one of the four resulting megaspores survives. This megaspore undergoes three rounds of mitotic division, resulting in seven cells with eight haploid nuclei (the central cell has two nuclei, called the polar nuclei). The lower end of the embryo sac consists of the haploid egg cell positioned in the middle of two other haploid cells, called synergids. The synergids function in the attraction and guidance of the pollen tube to the megagametophyte through the micropyle. At the upper end of the megagametophyte are three antipodal cells.

Microgametophyte

The male gametophytes, or microgametophytes, that participate in double fertilization are contained within pollen grains. They develop within the microsporangia, or pollen sacs, of the anthers on the stamens. Each microsporangium contains diploid microspore mother cells, or microsporocytes. Each microsporocyte undergoes meiosis, forming four haploid microspores, each of which can eventually develop into a pollen grain. A microspore undergoes mitosis and cytokinesis in order to produce two separate cells, the generative cell and the tube cell. These two cells in addition to the spore wall make up an immature pollen grain. As the male gametophyte matures, the generative cell passes into the tube cell, and the generative cell undergoes mitosis, producing two sperm cells. Once the pollen grain has matured, the anthers break open, releasing the pollen. The pollen is carried to the pistil of another flower, by wind or animal pollinators, and deposited on the stigma. As the pollen grain germinates, the tube cell produces the pollen tube, which elongates and extends down the long style of the carpel and into the ovary, where its sperm cells are released in the megagametophyte. Double fertilization proceeds from here.

Microgametogenesis

Microgametogenesis is the process in plant reproduction where a microgametophyte develops in a pollen grain to the three-celled stage of its development. In flowering plants it occurs with a microspore mother cell inside the anther of the plant.

When the microgametophyte is first formed inside the pollen grain four sets of fertile cells called sporogenous cells are apparent. These cells are surrounded by a wall of sterile cells called the tapetum, which supplies food to the cell and eventually becomes the cell wall for the pollen grain. These sets of sporogenous cells eventually develop into diploid microspore mother cells. These microspore mother cells, also called microsporocytes, then undergo meiosis and become four microspore haploid cells. These new microspore cells then undergo mitosis and form a tube cell and a generative cell. The generative cell then undergoes mitosis one more time to form two male gametes, also called sperm.

Dioecy

Dioecy (Greek "two households"; adjective form: dioecious) is a characteristic of a species, meaning that it has distinct male and female individual organisms or colonies, meaning that a colony contains only either male or female individuals. Dioecious reproduction is biparental reproduction. Dioecy is one method that excludes self-fertilization and promotes allogamy (outcrossing), and thus tends to reduce the expression of recessive deleterious mutations present in a population. Flowering plants have several other methods of excluding self-fertilization, called Self-incompatibility.

In Botany

Dioecious species have the male and female reproductive structures on separate plants. The meaning of *male* and *female* in the context of plants is different from that used in animal groups and the usage is not strictly correct. About 6 percent of angiosperm species are entirely dioecious and about 7% of angiosperm genera contain some dioecious species. Examples of dioecious plant species include ginkgos, willows, cannabis and African teak.

Dioecy occurs in a wide variety of plant families. However it is more common in woody plants and heterotrophic species.

Alternatives to Dioecy

There are several alternatives to dioecy for sexual reproductive structure organization in plants including bisexual flowers, monoecy, gynomonoecy, andromonecy. These are described at Plant reproductive morphology.

Alternation of Generations

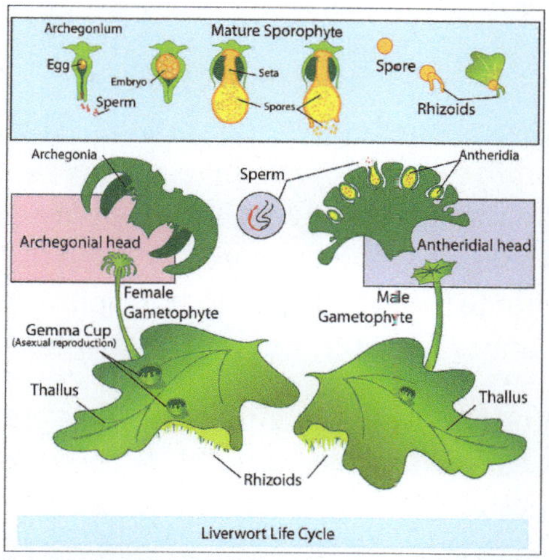

Liverwort species are either dioecious (dioicous) as shown here or monoecious (monoicous), with both antheridia and archegonia on the same gametophyte

Land plants have alternation of generations, for which the sporophyte generation produces spores rather than gametes. Strictly speaking, the sporophytes of land plants do not have either male or female reproductive organs. The gametophytes of flowering plants are of a single sex; the male gametophytes are contained within the pollen, and the female gametophytes are contained within ovules. The sporophyte generation is called dioecious when each sporophyte has only one kind of spore-producing organ whose spores ultimately give rise to either all male gametes (sperm) or all female gametes (eggs). Slightly different terms, "dioicous" and "monoicous", may be used for the gametophyte generation, although "dioecious" and "monoecious" are also used.

Monoicous

Monoicous plants are those species that bear both sperm and eggs on the same gametophyte. Dioicous plants are those that have gametophytes that produce only sperm or eggs but never both. The terms are used largely but not exclusively in the context of bryophytes. Both monoicous and dioicous gametophytes produce gametes in gametangia by mitosis rather than meiosis, so that sperm and eggs are genetically identical with their parent gametophyte. The states of being monoicous or dioicous are called monoicy and dioicy respectively.

Etymology and History

The word monoicous and the related forms mon(o)ecious are derived from the Greek (*mónos*), *single*, and (*oîkos*) or (*oikía*), *house*. The words dioicous and di(o)ecious are derived from (di-), twice, double. (*(o)e* is the Latin way of transliterating Greek οι, whereas *oi* is a more straightforward modern way.) Generally, the terms "monoicous" and "dioicous" have been restricted to description of haploid sexuality (gametophytic sexuality), and are thus used primarily to describe bryophytes in which the gametophyte is the dominant generation. Meanwhile, "monoecious" and "dioecious" are used to describe diploid sexuality (sporophytic sexuality), and thus are used to describe tracheophytes (vascular plants) in which the sporophyte is the dominant generation. However, this usage, although precise, is not universal, and "monoecious" and "dioecious" are still used by some bryologists for the gametophyte.

Bryophyte Sexuality

Bryophytes have life cycles that are gametophyte dominated. The longer lived, more prominent autotrophic plant is the gametophyte. The sporophyte in mosses and liverworts consists of an unbranched stalk (a seta) bearing a single sporangium or spore-producing capsule. Even when capable of photosynthesis, as in mosses and hornworts, bryophyte sporophytes require additional photosynthate from the gametophyte to sustain growth and spore development and are dependent on the gametophyte for their supplies of water, mineral nutrients and nitrogen.

Antheridia and archegonia are often clustered. A cluster of antheridia is called an androecium while a cluster of archegonia is called a gynoecium. (Note these terms have a different meaning when used to refer to flower structures.)

Bryophytes have the most elaborate gametophytes of all living land plants, and thus have a wide variety of gametangium positions and developmental patterns.

Gametangia are typically borne on the tips of shoots, but may also be found in the axils of leaves, under thalli or on elaborate structures called gametangiophores.

Bryophyte species may be:

- Autoicous meaning that androecia and gynoecia are found on the same individual (monoicous) but in distinctly separate locations. If these are on separate branches, the term cladautoicous can be applied.
- Synoicous (also called androgynous) bryophytes produce antheridia and archegonia interspersed in the same cluster.
- Paroicous bryophytes produce antheridia and archegonia in separate clusters in different leaf axils.

- Heteroicous bryophyte species may be either monoicous or sequentially dioicous depending on environmental conditions. This condition is also called polygamous or polyoicous.

Role in Survival

There can be both selective advantages and selective disadvantages for organisms that are monoicous or dioicous. Monoicous bryophytes can easily reproduce sexually, since both sexes can be found on the same organism. On the other hand, this can lead to inbreeding and reduce genetic variation within populations. Dioicous organisms necessarily exchange genes with other organisms of the species during sexual reproduction, increasing heterozygosity and variability (given a sufficiently large variable mating population). If isolated, however, organisms may only reproduce asexually, which could present a severe selective disadvantage over time. Bryophyte sperm dispersal can therefore be key to species longevity, particularly in dioicous species. While sperm dispersal is typically passive, with sperm dispersing through water, certain species exhibit very active dispersal mechanisms, such as aerial dispersal recently described in the liverwort *Conocephalum conicum*.

Stamen

Stamens of a *Hippeastrum* with white filaments and prominent anthers carrying pollen

The stamen (plural *stamina* or *stamens*) is the pollen-producing reproductive organ of a flower. Collectively the stamens form the androecium.

Morphology and Terminology

A stamen typically consists of a stalk called the filament and an anther which contains *microsporangia*. Most commonly anthers are two-lobed and are attached to the filament either at the base or in the middle area of the anther. The sterile tissue between the lobes is called the connective. A pollen grain develops from a microspore in the microsporangium and contains the male gametophyte.

The stamens in a flower are collectively called the *androecium*. The androecium can consist of as few a one-half stamen (i.e. a single locule) as in *Canna* species or as many as 3,482 stamens which have been counted in *Carnegiea gigantea*. The androecium in various species of plants forms a great variety of patterns, some of them highly complex. It surrounds the gynoecium and is surrounded by the perianth. A few members of the family Triuridaceae, particularly *Lacandonia schismatica*, are exceptional in that their gynoecia surround their androecia.

Hippeastrum flowers showing stamens above the style (with its terminal stigma)

Etymology

- *Stamen* is the Latin word meaning "thread" (originally thread of the warp, in weaving).
- Filament derives from classical Latin *filum*, meaning "thread".
- Anther derives from French *anthère*, from classical Latin *anthera*, meaning "medicine extracted from the flower" in turn from Ancient Greek.
- Androecium derives from Ancient Greek.

Variation in Morphology

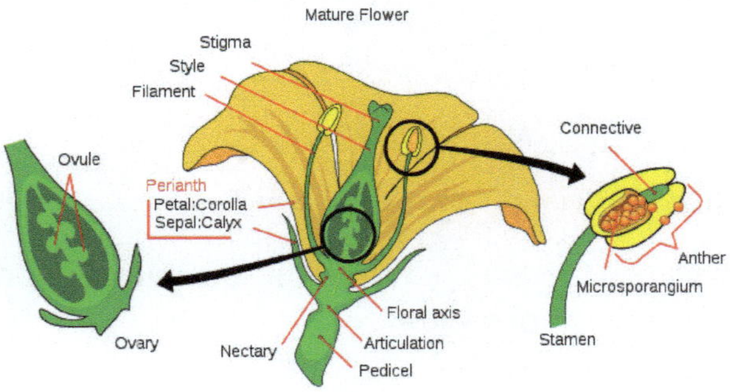

Stamens, with distal anther attached to the filament stalk, in context of floral anatomy

Depending on the species of plant, some or all of the stamens in a flower may be attached to the petals or to the floral axis. They also may be free-standing or fused to one another in many different ways, including fusion of some but not all stamens. The filaments may be fused and the anthers free, or the filaments free and the anthers fused. Rather than there being two locules, one locule of a stamen may fail to develop, or alternatively the two locules may merge late in development to give a single locule. Extreme cases of stamen fusion occur in some species of *Cyclanthera* in the family Cucurbitaceae and in section *Cyclanthera* of genus *Phyllanthus* (family Euphorbiaceae) where the stamens form a ring around the gynoecium, with a single locule.

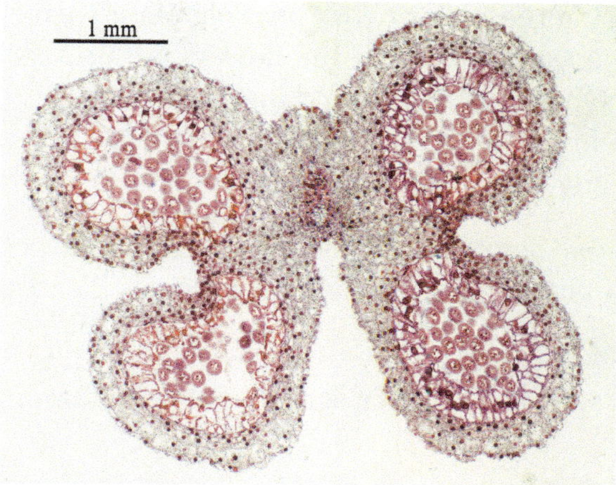

Cross section of a *Lilium* stamen, with four locules surrounded by the tapetum

Pollen Production

A typical anther contains four microsporangia. The *microsporangia* form sacs or pockets (*locules*) in the anther (anther sacs or pollen sacs). The two separate locules on each side of an anther may fuse into a single locule. Each microsporangium is lined with a nutritive tissue layer called the *tapetum* and initially contains diploid pollen mother cells. These undergo meiosis to form haploid spores. The spores may remain attached to each other in a tetrad or separate after meiosis. Each microspore then divides mitotically to form an immature microgametophyte called a pollen grain.

The pollen is eventually released when the anther forms openings (dehisces). These may consist of longitudinal slits, pores, as in the heath family (Ericaceae), or by valves, as in the barberry family (Berberidaceae). In some plants, notably members of Orchidaceae and Asclepiadoideae, the pollen remains in masses called pollinia, which are adapted to attach to particular pollinating agents such as birds or insects. More commonly, mature pollen grains separate and are dispensed by wind or water, pollinating insects, birds or other pollination vectors.

Pollen of angiosperms must be transported to the stigma, the receptive surface of the *carpel*, of a compatible flower, for successful pollination to occur. After arriving, the pollen grain (an immature microgametophyte) typically completes its development. It may grow a pollen tube and undergoing mitosis to produce two sperm nuclei.

Sexual Reproduction in Plants

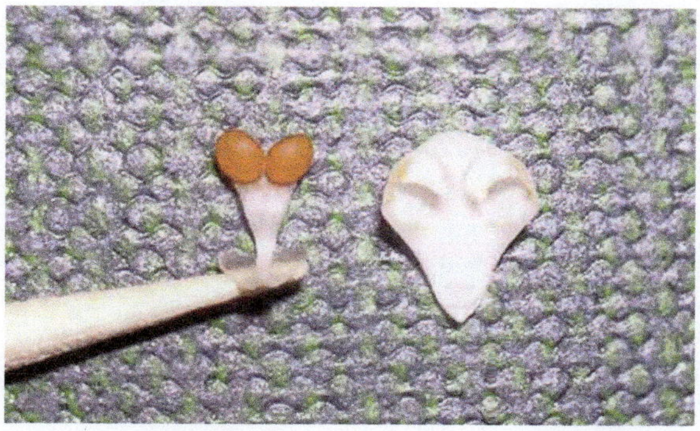
Stamen with pollinia and its anther cap. *Phalaenopsis* orchid.

In the typical flower (that is, in the majority of flowering plant species) each flower has both carpels and stamens. In some species, however, the flowers are unisexual with only carpels or stamens. (monoecious = both types of flowers found on the same plant; dioecious = the two types of flower found only on different plants). A flower with only stamens is called androecious. A flower with only carpels is called gynoecious.

A flower having only functional stamens and lacking functional carpels is called a staminate flower, or (inaccurately) male. A plant with only functional carpels is called pistillate, or (inaccurately) female.

An abortive or rudimentary stamen is called a staminodium or staminode, such as in *Scrophularia nodosa*.

The carpels and stamens of orchids are fused into a column. The top part of the column is formed by the anther, which is covered by an anther cap.

Descriptive Terms

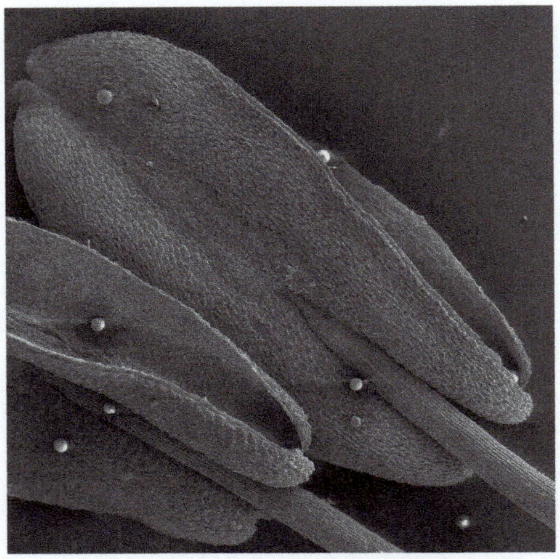

Scanning electron microscope image of *Pentas lanceolata* anthers, with pollen grains on surface

Lily stamens with prominent red anthers and white filaments

- A column formed from the fusion of multiple filaments is known as an *androphore*.

The anther can be attached to the filament's connective in two ways:

- basifixed: attached at its base to the filament
    - pseudobasifixed: a somewhat misnomer configuration where connective tissue extends in a tube around the filament
- dorsifixed: attached at its center to the filament, usually versatile (able to move)

Stamens can be connate (fused or joined in the same whorl):

- extrorse: anther dehiscence directed away from the centre of the flower. Cf. introrse, directed inwards, and latrorse towards the side.
- monadelphous: fused into a single, compound structure
- declinate: curving downwards, then up at the tip (also - declinate-descending)
- diadelphous: joined partially into two androecial structures
- pentadelphous: joined partially into five androecial structures
- synandrous: only the anthers are connate (such as in the Asteraceae). The fused stamens are referred to as a synandrium.

Stamens can also be adnate (fused or joined from more than one whorl):

- epipetalous: adnate to the corolla
- epiphyllous: adnate to undifferentiated tepals (as in many Liliaceae)

They can have different lengths from each other:

- didymous: two equal pairs
- didynamous: occurring in two pairs, a long pair and a shorter pair
- tetradynamous: occurring as a set of six stamens with four long and two shorter ones

or respective to the rest of the flower (perianth):

- exserted: extending beyond the corolla
- included: not extending beyond the corolla

They may be arranged in one of two different patterns:

- spiral; or
- whorled: one or more discrete whorls (series)

They may be arranged, with respect to the petals:

- diplostemonous: in two whorls, the outer alternating with the petals, while the inner is opposite the petals
- obdiplostemonous: in two whorls, the outer opposite the petals.

Stigma (Botany)

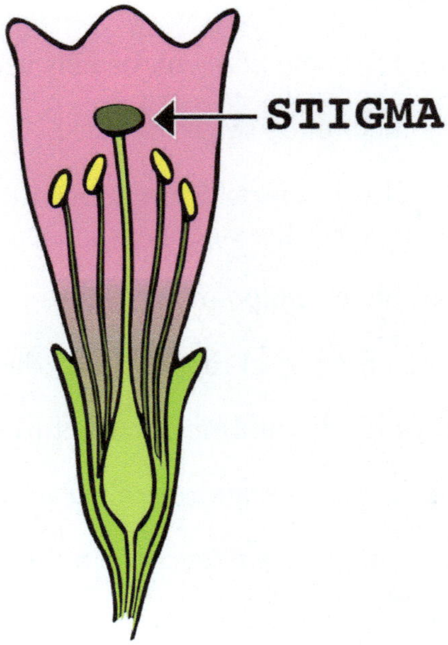

Diagram of stigma

The stigma (plural: stigmata) is the receptive tip of a carpel, or of several fused carpels, in the gynoecium of a flower.

Description

Stigma of a *Tulipa* species, with pollen

The stigma, together with the style and ovary comprises the pistil, which in turn is part of the gynoecium or female reproductive organ of a plant. The stigma forms the distal portion of the style or stylodia. The stigma is composed of stigmatic papillae, the cells which are receptive to pollen. These may be restricted to the apex of the style or, especially in wind pollinated species, cover a wide surface.

Corn stigma called "silk"

The stigma receives pollen and it is on the stigma that the pollen grain germinates. Often sticky, the stigma is adapted in various ways to catch and trap pollen with various hairs, flaps, or sculpturings. The pollen may be captured from the air (wind-borne pollen, anemophily), from visiting insects or other animals (biotic pollination), or in rare cases from surrounding water (hydrophily). Stigma can vary from long and slender to globe shaped to feathery.

Pollen is typically highly desiccated when it leaves an anther. Stigma have been shown to assist in the rehydration of pollen and in promoting germination of the pollen tube. Stigma also ensure proper adhesion of the correct species of pollen. Stigma can play an active role in pollen discrimination and some self-incompatibility reactions, that reject pollen from the same or genetically similar plants, involve interaction between the stigma and the surface of the pollen grain.

Shape

The stigma is often split into lobes, e.g. trifid (three lobed), and may resemble the head of a pin (capitate), or come to a point (punctiform). The shape of the stigma may vary considerably:

Stigma Shapes

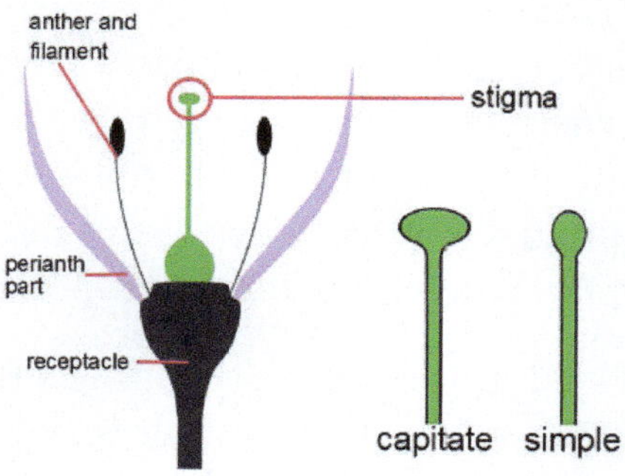

Capitate and simple

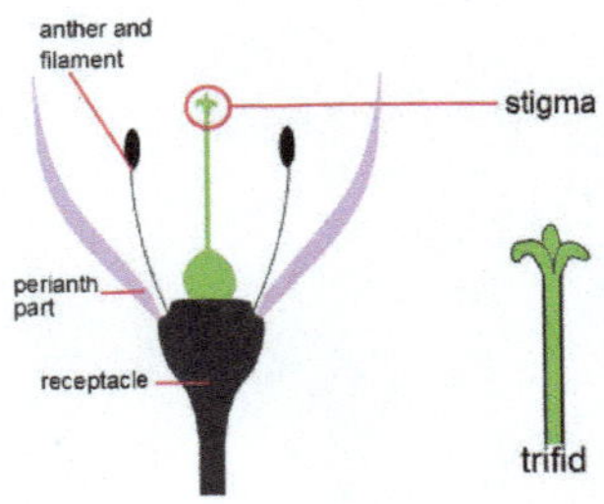

Trifid

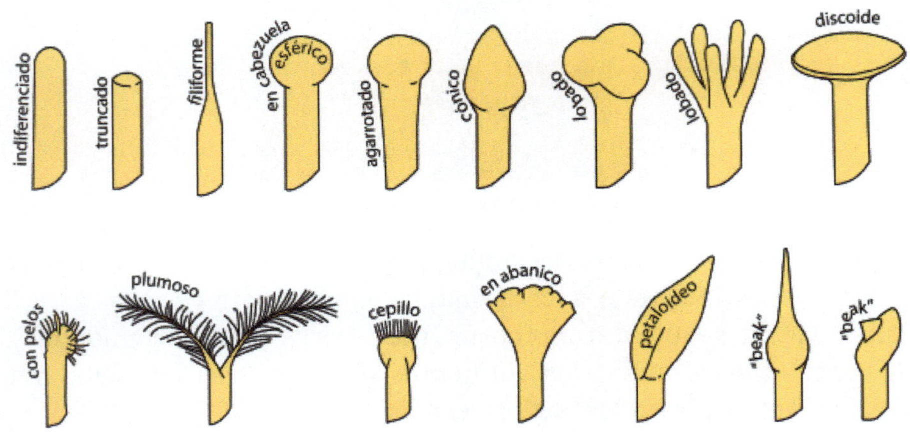

Style

Structure

The style is a narrow upward extension of the ovary, connecting it to the stigmatic papillae. It may be absent in some plants in the case the stigma is referred to as sessile. Styles are generally tube-like — either long or short. The style can be open (containing few or no cells in the central portion) with a central canal which may be filled with mucilage. Alternatively the style may be closed (densely packed with cells throughout). Most syncarpous monocots and some eudicots have open styles, while many syncarpous eudicots and grasses have closed (solid) styles containing specialised secretory transmitting tissue tissue, linking the stigma to the centre of the ovary. This forms a nutrient rich tract for pollen tube growth.

Where there are more than one carpels to the pistil, each may have a separate style-like stylodium, or share a common style. In Irises and others in the Iridaceae family, the style divides into three petal-like (petaloid), *style branches* (sometimes also referred to as stylodia), almost to the base of the style and is called tribrachiate. These are flaps of tissue, running from the perianth tube above the sepal. The stigma is a rim or edge on the underside of the branch, near the end lobes. Style branches also appear on *Dietes*, *Pardanthopsis* and most species of *Moraea*.

In *Crocuses*, there are three divided style branches, creating a tube. *Hesperantha* has a spreading style branch. Alternatively the style may be lobed rather than branched. *Gladiolus* has a bi-lobed style branch (bilobate). *Freesia*, *Lapeirousia*, *Romulea*, *Savannosiphon* and *Watsonia* have bifuracated (two branched) and recurved style branches.

Style Morphology

Iris versicolor showing three structures with two overlapping lips, an upper petaloid style branch and a lower tepal, enclosing a stamen

Iris missouriensis showing the pale blue style branch above the drooping petal

The feathery stigma of *Crocus sativus* has branches corresponding to three carpels

Attachment to the Ovary

Style Position

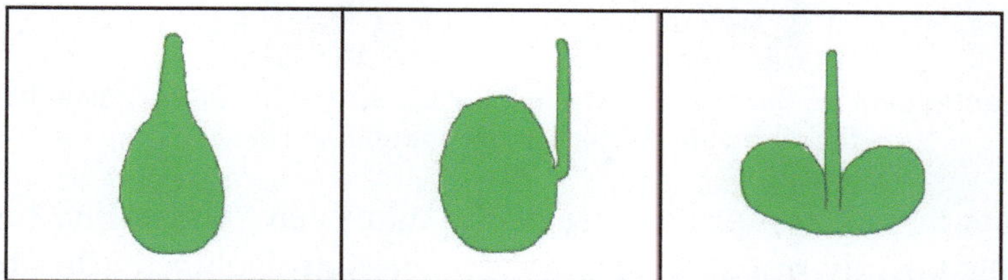

Terminal (apical)	Lateral	Gynobasic

May be terminal (apical), subapical, lateral, gynobasic, or subgynobasic. Terminal (apical) style position refers to attachment at the apex of the ovary and is the commonest pattern. In the subapical pattern the style arises to the side slightly below the apex. a lateral style arises from the side of the ovary and is found in Rosaceae. The gynobasic style arises from the base of the ovary, or between the ovary lobes and is characteristic of Boraginaceae. Subgynobasic styles characterise *Allium*.

Pollination

Pollen tubes grow the length of the style to reach the ovules, and in some cases self-incompatibility reactions in the style prevent full growth of the pollen tubes. In some species, including *Gasteria* at least, the pollen tube is directed to the micropyle of the ovule by the style.

Vegetative Reproduction

Vegetative reproduction (vegetative propagation, vegetative multiplication, vegetative cloning) is a form of asexual reproduction in plants. It is a process by which new organisms arise without production of seeds or spores. It can occur naturally or be induced by horticulturists.

Production of new individuals along a leaf margin of the air plant, *Kalanchoe pinnata*. The small plant in front is about 1 cm tall. The concept of "individual" is obviously stretched by this process.

Bryophyllum daigremontianum produces plantlets along the margins of its leaves. When they are mature enough, they drop off and root in any suitable soil beneath.

Although most plants normally reproduce sexually, many have the ability for vegetative propagation, or can be vegetatively propagated if small pieces are subjected to chemical (hormonal) treatments. This is because meristematic cells capable of cellular differentiation are present in many plant tissues. Horticulturalists are interested in understanding how meristematic cells can be induced to reproduce an entire plant.

Vegetative reproduction from a stem cutting less than a week old. Some species are more conducive to this means of propagation than others.

A bulb of *Muscari* has reproduced vegetatively underground to make two bulbs, each of which produces a flower stem.

Success rates and difficulty of propagation vary greatly. For example, willow and coleus can be propagated merely by inserting a stem in water or moist soil. On the other hand, monocotyledons, unlike dicotyledons, typically lack a vascular cambium and therefore are harder to propagate.

Types

In a wide sense, methods of vegetative propagation include cutting, vegetative apomixis, layering, division, budding, grafting and tissue culture. Cutting is the most common artificial vegetative propagation method, where pieces of the "parent" plant are removed and placed in a suitable environment so that they can grow into a whole new plant, the "clone", which is genetically identical to the parent. Cutting exploits the ability of plants to grow adventitious roots (i.e. root material that can generate from a location other than the existing or primary root system, as in from a leaf or cut stem) under certain conditions. Vegetative propagation is usually considered a cloning method. However, there are several cases where vegetatively propagated plants are not genetically identical. Root cuttings of thornless blackberries will revert to thorny type because the adventitious shoot develops from a cell that is genetically thorny. Thornless blackberry is a chimera, with the epidermal layers genetically thornless but the tissue beneath it genetically thorny. Similarly, leaf cutting propagation of certain chimeral variegated plants, such as snake plant (Sansevieria trifasciata), will produce mainly nonvariegated plants. Grafting is often not a complete cloning method because seedlings are used as rootstocks. In that case only the top of the plant is clonal. In some crops, particularly apples, the rootstocks are vegetatively propagated so the entire graft can be clonal if the scion and rootstock are both clones. Apomixis (including apospory and diplospory) is a type of reproduction that does not involve fertilisation. In flowering plants, unfertilized seeds are involved, or plantlets that grow instead of flowers. Hawkweed (*Hieracium*), dandelion (*Taraxacum*), some citrus (*Citrus*) and many grasses such as Kentucky blue grass (*Poa pratensis*) all use this form of asexual reproduction. Bulbils are sometimes formed instead of the flowers of garlic.

Vegetative Structures

Virtually all types of shoots and roots are capable of vegetative propagation, including stems, basal shoots, tubers, rhizomes, stolons, corms, bulbs, and buds. In a few species (such as *Kalanchoë*), leaves are involved in vegetative reproduction.

- The rhizome is a modified underground stem serving as an organ of vegetative reproduction, e. g. *Polypodium* (polypody), iris, couch grass and nettles.
- Prostrate aerial stems, called runners or stolons are important vegetative reproduction organs in some species, such as the strawberry, numerous grasses, and some ferns.
- Adventitious buds form on roots near the ground surface, on damaged stems (as on the stumps of cut trees), or on old roots. These develop into above-ground stems and leaves.
- A form of budding called suckering is the reproduction or regeneration of a plant by shoots that arise from an existing root system. Species that characteristically produce suckers include Elm (*Ulmus*) and members of the Rose Family (*Rosa*).
- Another type of a vegetative reproduction is the production of bulbs. Plants like onion (*Allium cepa*), hyacinth (*Hyacinth*), narcissus (*Narcissus*) and tulips (*Tulipa*) reproduce by forming bulbs.
- Other plants like potatoes (*Solanum tuberosum*) and dahlia (*Dahlia*) reproduce by a method similar to bulbs: they produce tubers.
- Gladioli and crocuses (*Crocus*) reproduce by forming a bulb-like structure called a corm.
- Some orchids reproduce by the growth of keikis from the stem or cane of the parent plant.

Natural Vegetative Propagation

Natural vegetative propagation is mostly a process found in herbaceous and woody perennial plants, and typically involves structural modifications of the stem, although any horizontal, underground part of a plant (whether stem, leaf, or root) can contribute to vegetative reproduction of a plant. Most plant species that survive and significantly expand by vegetative reproduction would be perennial almost by definition, since specialized organs of vegetative reproduction, like seeds of annuals, serve to survive seasonally harsh conditions. A plant that persists in a location through vegetative reproduction of individuals over a long period of time constitutes a clonal colony.

In a sense, this process is not one of reproduction but one of survival and expansion of biomass of the individual. When an individual organism increases in size via cell multiplication and remains intact, the process is called "vegetative growth". However, in vegetative reproduction, the new plants that result are new individuals in almost every respect except genetic. Of considerable interest is how this process appears to reset the aging clock.

Artificial Vegetative Propagation

Mass Propagation of Eucalyptus Seedlings

Vegetative propagation of particular cultivars that have desirable characteristics is very common practice. Reasons for preferring vegetative rather than sexual means of reproduction vary, but commonly include greater ease and speed of propagation of certain plants, such as many perennial root crops and vines. Another major attraction is that the resulting plant amounts to a clone of the parent plant and accordingly is of a more predictable quality than most seedlings. However, as can be seen in many variegated plants, this does not always apply, because many plants actually are chimeras and cuttings might reflect the attributes of only one or some of the parent cell lines. Man-made methods of vegetative reproduction are usually enhancements of natural processes, but they range from rooting cuttings to grafting and artificial propagation by laboratory tissue culture.

In horticulture, a "cutting" is a piece that has been cut off from a mother plant and then caused to grow into a whole plant. Often this involves a piece of stem that is treated with rooting liquid or powder containing hormones. In some species root cuttings can produce shoot growth. When the cutting has become a self-sufficient plant, it is genetically identical to the mother plant except when chimeric tissues or similar complications affect the outcome.

A related form of regeneration is that of grafting. A stem piece or a single bud (the scion) is joined onto the stem of a plant that has roots (the rootstock), or a stem piece can be joined to a root piece. A popular use of grafting is to produce fruit trees, sometimes with more than one variety of the same fruit species growing from the same stem. Rootstocks for fruit trees are either seedlings or propagated by layering.

Cultivated Plants Propagated by Vegetative Methods

A number of commonly cultivated plants are usually propagated by vegetative means rather than by seeds. This is a listing of such plants:

African violets — leaf cuttings

Apple — grafting

Avocado — grafting

Banana — sucker removal

Blackberries (*Rubus occidentalis*) — stem cuttings

Bryophyllum — leaf cuttings

Canna — division

Cannabis — stem cuttings

Citrus (lemon, orange, grapefruit, Tangerine, dayap) — grafting

Date — sucker removal

Fig — stem cuttings

Garden strawberry — runners (stolons)

Grapes — stem cuttings, grafting, aerial layering

Hops — stem cuttings

Jasmine - stem cuttings

Kumquat - stem cuttings, grafting

Manioc (cassava) — stem cuttings

Maple — stem cuttings, grafting

Nut crops (walnut, pecan) — grafting

Olive — stem cuttings and layering

Peach — grafting

Pear — grafting

Pineapple — stem cuttings

Plum — stem cuttings

Poplar — stem cuttings

Potato — stem (tuber) cuttings

Rose — stem cuttings

Sansevieria - Leaf cuttings or division

Sugar cane — stem cuttings

Tea — stem cuttings

Vanilla — stem cuttings

Verbena — stem cuttings

Willow — stem cuttings

References

- Renner, S.S. & Ricklefs, R.E. (1995). "Dioecy and its correlates in the flowering plants". American Journal of Botany. 82: 596–606. doi:10.2307/2445418. JSTOR 2445418
- Beentje, Henk (2010). The Kew Plant Glossary. Richmond, Surrey: Royal Botanic Gardens, Kew. ISBN 978-1-84246-422-9
- Whittemore, Alan T. "Ranunculus". http://www.efloras.org/florataxon.aspx?flora_id=1&taxon_id=127971. Retrieved 2013-03-04
- Barrett, S.C.H. (2002). "The evolution of plant sexual diversity" (PDF). Nature Reviews Genetics. 3 (4): 274–284. doi:10.1038/nrg776
- Buzgo, Matyas; Soltis, Pamela S. & Soltis, Douglas E. (2004). "Floral Developmental Morphology of Amborella trichopoda (Amborellaceae)". International Journal of Plant Sciences. 165 (6): 925–947. doi:10.1086/424024
- David Sadava; H. Craig Heller; Gordon H. Orians; William K. Purves; David M. Hillis (December 2006). Life: the science of biology. Macmillan. pp. 477–. ISBN 978-0-7167-7674-1. Retrieved 4 August 2010
- "As easy as APG III - Scientists revise the system of classifying flowering plants". The Linnean Society of London. 2009-10-08. Retrieved 2009-10-02
- Edwards, D (2000). "The role of mid-palaeozoic mesofossils in the detection of early bryophytes". Philos Trans R Soc Lond B Biol Sci. 355 (1398): 733–755. doi:10.1098/rstb.2000.0613. PMC 1692787. PMID 10905607
- Stace, H.M. (1995). "Protogyny, Self-Incompatibility and Pollination in Anthocercis gracilis (Solanaceae)". Australian Journal of Botany. 43 (5): 451–459. doi:10.1071/BT9950451

- Stewart, Wilson Nichols; Rothwell, Gar W. (1993). Paleobotany and the evolution of plants (2nd ed.). Cambridge Univ. Press. p. 498. ISBN 0-521-23315-1
- V. Raghavan (September 2003). "Some reflections on double fertilization, from its discovery to the present". New Phytologist. 159 (3): 565–583. doi:10.1046/j.1469-8137.2003.00846.x. Retrieved 2013-08-27
- Olson, Matthew S. & Antonovics, Janis (2000). "Correlation between male and female reproduction in the subdioecious herb Astilbe biternata (Saxifragaceae)". American Journal of Botany. 87 (6): 837. doi:10.2307/2656891
- Buchmann, Stephen L.; Nabhan, Gary Paul (2012). The Forgotten Pollinators. Island Press. pp. 41–42. ISBN 978-1-59726-908-7
- Scotland, R. W.; Wortley, A. H. (2003). "How many species of seed plants are there?". Taxon. 52 (1): 101–104. doi:10.2307/3647306. JSTOR 3647306
- Soltis, D. E.; Soltis, P. S. (2004). "Amborella not a "basal angiosperm"? Not so fast". American Journal of Botany. 91 (6): 997–1001. doi:10.3732/ajb.91.6.997. PMID 21653455
- Raven, Peter H., Ray F. Evert, & Susan E. Eichhorn, 2005. Biology of Plants, 7th edition. (New York: W. H. Freeman and Company). ISBN 0-7167-1007-2
- Govaerts, R. (2003). "How many species of seed plants are there? – a response". Taxon. 52 (3): 583–584. doi:10.2307/3647457. JSTOR 3647457
- Jeffrey D. Palmer; Douglas E. Soltis; Mark W. Chase (2004). "The plant tree of life: an overview and some points of view". American Journal of Botany. 91 (10): 1437–1445. doi:10.3732/ajb.91.10.1437. PMID 21652302
- Raghavan, V. (2005). Double fertilization: embryo and endosperm development in flowering plants (illustrated ed.). Birkhäuser. pp. 17–19. ISBN 3-540-27791-9
- K. S. Bawa (1980). "Evolution of Dioecy in Flowering Plants". Annual Review of Ecology and Systematics. 11: 15–39. doi:10.1146/annurev.es.11.110180.000311
- Matallana, G.; Wendt, T.; Araujo, D.S.D.; Scarano, F.R. (2005), "High abundance of dioecious plants in a tropical coastal vegetation", American Journal of Botany, 92 (9): 1513–1519, doi:10.3732/ajb.92.9.1513
- Walters, Dirk R Walters Bonnie By (1996). Vascular plant taxonomy. Dubuque, Iowa: Kendall/Hunt Pub. Co. p. 124. ISBN 978-0-7872-2108-9
- Nickrent D.L., Musselman L.J. (2004). "Introduction to Parasitic Flowering Plants". The Plant Health Instructor. doi:10.1094/PHI-I-2004-0330-01
- Shimamura, Masaki; Yamaguchi, Tomio; Deguchi, Hironori (2007). "Airborne sperm of Conocephalum conicum (Conocephalaceae)". Journal of Plant Research. 121 (1): 69–71. doi:10.1007/s10265-007-0128-6

2

Seed and Germination Process: A Comprehensive Study

A seed usually consists of an embryo and a seed coat. Seeds store essential nutrients required for the growth of the plant. Seed dispersal, seedling and cotyledon are natural methods of plant propagation. The chapter on seed and germination process offers an insightful focus, keeping in mind the complex subject matter.

Seed

A seed is an embryonic plant enclosed in a protective outer covering. The formation of the seed is part of the process of reproduction in seed plants, the spermatophytes, including the gymnosperm and angiosperm plants.

Brown flax seeds

Seeds are the product of the ripened ovule, after fertilization by pollen and some growth within the mother plant. The embryo is developed from the zygote and the seed coat from the integuments of the ovule.

Seeds have been an important development in the reproduction and success of gymnosperms and angiosperms plants, relative to more primitive plants such as ferns, mosses and liverworts, which do not have seeds and use water-dependant means to propagate themselves. Seed plants now dominate biological niches on land, from forests to grasslands both in hot and cold climates.

The term “seed” also has a general meaning that antedates the above—anything that can be sown, e.g. “seed” potatoes, “seeds” of corn or sunflower “seeds”. In the case of sunflower and corn “seeds”, what is sown is the seed enclosed in a shell or husk, whereas the potato is a tuber.

Many structures commonly referred to as “seeds” are actually dry fruits. Plants producing berries are called baccate. Sunflower seeds are sometimes sold commercially while still enclosed within the hard wall of the fruit, which must be split open to reach the seed. Different groups of plants have other modifications, the so-called stone fruits (such as the peach) have a hardened fruit layer (the endocarp) fused to and surrounding the actual seed. Nuts are the one-seeded, hard-shelled fruit of some plants with an indehiscent seed, such as an acorn or hazelnut.

Production

Seeds are produced in several related groups of plants, and their manner of production distinguishes the angiosperms ("enclosed seeds") from the gymnosperms ("naked seeds"). Angiosperm seeds are produced in a hard or fleshy structure called a fruit that encloses the seeds, hence the name. Some fruits have layers of both hard and fleshy material. In gymnosperms, no special structure develops to enclose the seeds, which begin their development "naked" on the bracts of cones. However, the seeds do become covered by the cone scales as they develop in some species of conifer.

Seed production in natural plant populations varies widely from year to year in response to weather variables, insects and diseases, and internal cycles within the plants themselves. Over a 20-year period, for example, forests composed of loblolly pine and shortleaf pine produced from 0 to nearly 5 million sound pine seeds per hectare. Over this period, there were six bumper, five poor, and nine good seed crops, when evaluated in regard to producing adequate seedlings for natural forest reproduction.

Development

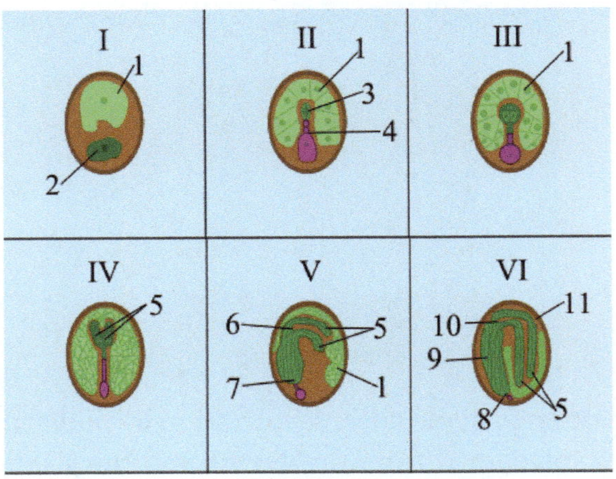

Stages of Seed Development

I Zygote	IV Heart
II Proembryo	V Torpedo
III Globular	VI Mature Embryo

Key: 1. *Endosperm* 2. *Zygote* 3. *Embryo* 4. *Suspensor* 5. *Cotyledons* 6. *Shoot Apical Meristem* 7. *Root Apical Meristem* 8. *Radicle* 9. *Hypocotyl* 10. *Epicotyl* 11. *Seed Coat*

Angiosperm (flowering plants) seeds consist of three genetically distinct constituents: (1) the embryo formed from the zygote, (2) the endosperm, which is normally triploid, (3) the seed coat from tissue derived from the maternal tissue of the ovule. In angiosperms, the process of seed development begins with double fertilization, which involves the fusion of two male gametes with the egg cell and the central cell to form the primary endosperm and the zygote. Right after fertilization, the zygote is mostly inactive, but the primary endosperm divides rapidly to form the endosperm tissue. This tissue becomes the food the young plant will consume until the roots have developed after germination.

Ovule

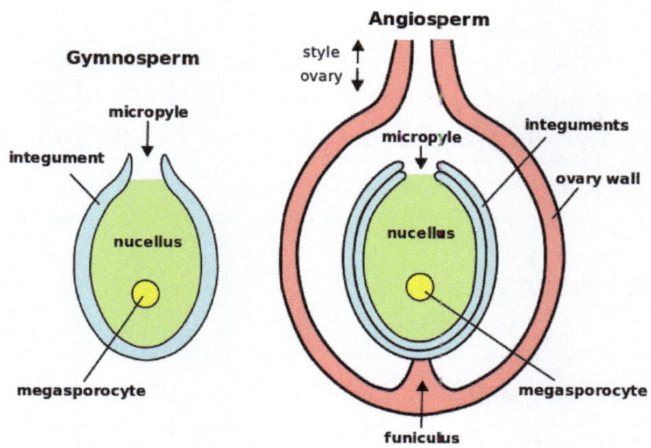

Plant ovules: Gymnosperm ovule on left, angiosperm ovule (inside ovary) on right

After fertilization the ovules develop into the seeds. The ovule consists of a number of components:

- The funicle (*funiculus, funiculi*) or seed stalk which attaches the ovule to the placenta and hence ovary or fruit wall, at the pericarp.
- The nucellus, the remnant of the megasporangium and main region of the ovule where the megagametophyte develops.
- The micropyle, a small pore or opening in the apex of the integument of the ovule where the pollen tube usually enters during the process of fertilization.
- The chalaza, the base of the ovule opposite the micropyle, where integument and nucellus are joined together).

The shape of the ovules as they develop often affects the final shape of the seeds. Plants generally produce ovules of four shapes: the most common shape is called anatropous, with a curved shape. Orthotropous ovules are straight with all the parts of the ovule lined up in a long row producing an uncurved seed. Campylotropous ovules have a curved megagametophyte often giving the seed a tight “C” shape. The last ovule shape is called amphitropous, where the ovule is partly inverted and turned back 90 degrees on its stalk (the funicle or funiculus).

In the majority of flowering plants, the zygote’s first division is transversely oriented in regards to the long axis, and this establishes the polarity of the embryo. The upper or chalazal pole becomes the main area of growth of the embryo, while the lower or micropylar pole produces the stalk-like suspensor that attaches to the micropyle. The suspensor absorbs and manufacturers nutrients from the endosperm that are used during the embryo’s growth.

Embryo

The inside of a *Ginkgo* seed, showing a well-developed embryo, nutritive tissue (megagametophyte), and a bit of the surrounding seed coat

The main components of the embryo are:

- The cotyledons, the seed leaves, attached to the embryonic axis. There may be one (Monocotyledons), or two (Dicotyledons). The cotyledons are also the source of nutrients in the non-endospermic dicotyledons, in which case they replace the endosperm, and are thick and leathery. In endospermic seeds the cotyledons are thin and papery. Dicotyledons have the point of attachment opposite one another on the axis.
- The epicotyl, the embryonic axis above the point of attachment of the cotyledon(s).
- The plumule, the tip of the epicotyl, and has a feathery appearance due to the presence of young leaf primordia at the apex, and will become the shoot upon germination.
- The hypocotyl, the embryonic axis below the point of attachment of the cotyledon(s), connecting the epicotyle and the radicle, being the stem-root transition zone.
- The radicle, the basal tip of the hypocotyl, grows into the primary root.

Monocotyledonous plants have two additional structures in the form of sheaths. The plumule is covered with a coleoptile that forms the first leaf while the radicle is covered with a coleorhiza that connects to the primary root and adventitious roots form from the sides. Here the hypocotyl is a rudimentary axis between radicle and plumule. The seeds of corn are constructed with these structures; pericarp, scutellum (single large cotyledon) that absorbs nutrients from the endosperm, plumule, radicle, coleoptile and coleorhiza—these last two structures are sheath-like and enclose the plumule and radicle, acting as a protective covering.

Seed Coat

The maturing ovule undergoes marked changes in the integuments, generally a reduction and disorganisation but occasionally a thickening. The seed coat forms from the two integuments or outer layers of cells of the ovule, which derive from tissue from the mother plant, the inner integument forms the tegmen and the outer forms the testa. (The seed coats of some mononocotyledon plants, such as the grasses, are not distinct structures, but are fused with the fruit wall to form a pericarp.)

The testae of both monocots and dicots are often marked with patterns and textured markings, or have wings or tufts of hair. When the seed coat forms from only one layer, it is also called the testa, though not all such testae are homologous from one species to the next. The funiculus abscises (detaches at fixed point – abscission zone), the scar forming an oval depression, the hilum. Anatropous ovules have a portion of the funiculus that is adnate (fused to the seed coat), and which forms a longitudinal ridge, or raphe, just above the hilum. In bitegmic ovules (e.g. *Gossypium* described here) both inner and outer integuments contribute to the seed coat formation. With continuing maturation the cells enlarge in the outer integument. While the inner epidermis may remain a single layer, it may also divide to produce two to three layers and accumulates starch, and is referred to as the colourless layer. By contrast the outer epidermis becomes tanniferous. The inner integument may consist of eight to fifteen layers. (Kozlowski 1972).

As the cells enlarge, and starch is deposited in the outer layers of the pigmented zone below the outer epidermis, this zone begins to lignify, while the cells of the outer epidermis enlarge radially and their walls thicken, with nucleus and cytoplasm compressed into the outer layer. these cells which are broader on their inner surface are called palisade cells. In the inner epidermis the cells also enlarge radially with plate like thickening of the walls. The mature inner integument has a palisade layer, a pigmented zone with 15-20 layers, while the innermost layer is known as the fringe layer. (Kozlowski 1972).

Gymnosperms

In gymnosperms, which do not form ovaries, the ovules and hence the seeds are exposed. This is the basis for their nomenclature – naked seeded plants. Two sperm cells transferred from the pollen do not develop the seed by double fertilization, but one sperm nucleus unites with the egg nucleus and the other sperm is not used.

Sometimes each sperm fertilizes an egg cell and one zygote is then aborted or absorbed during early development. The seed is composed of the embryo (the result of fertilization) and tissue from the mother plant, which also form a cone around the seed in coniferous plants such as pine and spruce.

Shape and Appearance

A large number of terms are used to describe seed shapes, many of which are largely self-explanatory such as *Bean-shaped* (reniform) – resembling a kidney, with lobed ends on either side of the hilum, *Square* or *Oblong* – angular with all sides more or less equal or longer than wide, *Triangular* – three sided, broadest below middle, *Elliptic* or *Ovate* or *Obovate* – rounded at both ends, or egg shaped (ovate or obovate, broader at one end), being rounded but either symmetrical about the middle or broader below the middle or broader above the middle.

Other less obvious terms include discoid (resembling a disc or plate, having both thickness and parallel faces and with a rounded margin), ellipsoid, globose (spherical), or subglobose (Inflated, but less than spherical), lenticular, oblong, ovoid, reniform and sectoroid. Striate seeds are striped with parallel, longitudinal lines or ridges. The commonest colours are brown and black, other colours are infrequent. The surface varies from highly polished to considerably roughened. The surface may have a variety of appendages. A seed coat with the consistency of cork is referred to as suberose. Other terms include crustaceous (hard, thin or brittle).

Structure

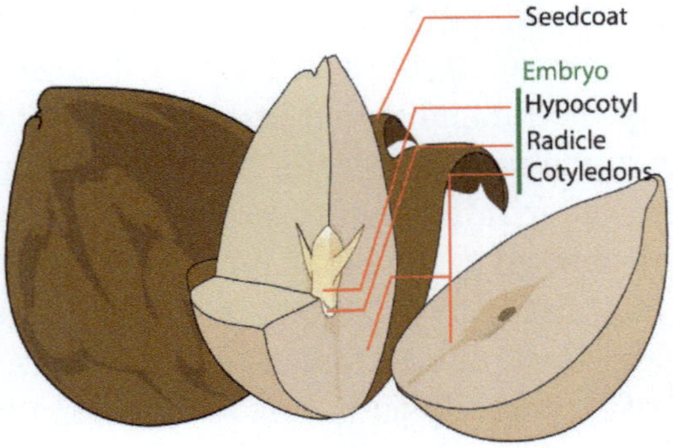

The parts of an avocado seed (a dicot), showing the seed coat and embryo

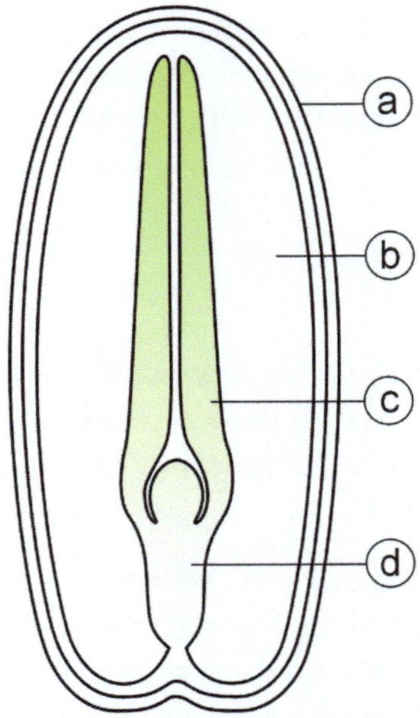

Diagram of the internal structure of a dicot seed and embryo:
(a) seed coat, (b) endosperm, (c) cotyledon, (d) hypocotyl

A typical seed includes two basic parts:

1. an embryo;
2. a seed coat.

In addition, the endosperm forms a supply of nutrients for the embryo in most monocotyledons and the endospermic dicotyledons.

Seed Types

Seeds have been considered to occur in many structurally different types (Martin 1946). These are based on a number of criteria, of which the dominant one is the embryo-to-seed size ratio. This reflects the degree to which the developing cotyledons absorb the nutrients of the endosperm, and thus obliterate it.

Six types occur amongst the monocotyledons, ten in the dicotyledons, and two in the gymnosperms (linear and spatulate). This classification is based on three characteristics: embryo morphology, amount of endosperm and the position of the embryo relative to the endosperm.

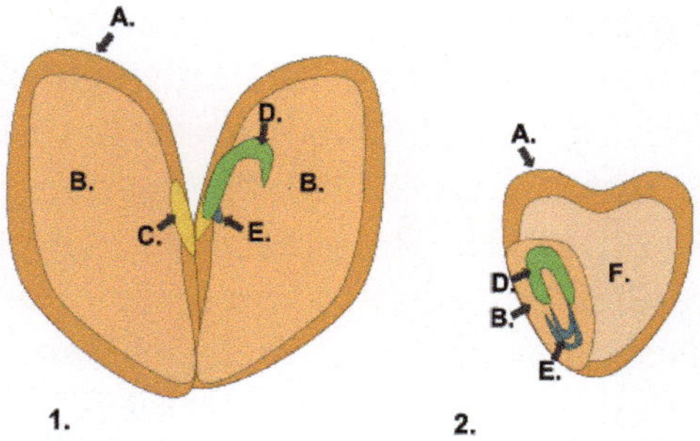

Diagram of a generalized dicot seed (1) versus a generalized monocot seed (2). A. Seed Coat B. Cotyledon C. Hilum D. Plumule E. Radicle F. Endosperm

Embryo

In endospermic seeds, there are two distinct regions inside the seed coat, an upper and larger endosperm and a lower smaller embryo. The embryo is the fertilised ovule, an immature plant from which a new plant will grow under proper conditions. The embryo has one cotyledon or seed leaf in monocotyledons, two cotyledons in almost all dicotyledons and two or more in gymnosperms. In the fruit of grains (caryopses) the single monocotyledon is shield shaped and hence called a scutellum. The scutellum is pressed closely against the endosperm from which it absorbs food, and passes it to the growing parts. Embryo descriptors include small, straight, bent, curved and curled.

Nutrient Storage

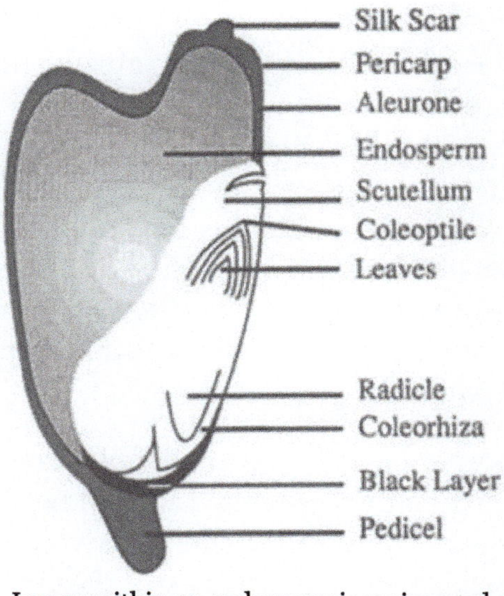

Layers within an endospermic maize seed

Within the seed, there usually is a store of nutrients for the seedling that will grow from the embryo. The form of the stored nutrition varies depending on the kind of plant. In angiosperms, the stored food begins as a tissue called the endosperm, which is derived from the mother plant and the pollen via double fertilization. It is usually triploid, and is rich in oil or starch, and protein.

In gymnosperms, such as conifers, the food storage tissue (also called endosperm) is part of the female gametophyte, a haploid tissue. The endosperm is surrounded by the aleurone layer (peripheral endosperm), filled with proteinaceous aleurone grains.

Originally, by analogy with the animal ovum, the outer nucellus layer (perisperm) was referred to as albumen, and the inner endosperm layer as vitellus. Although misleading, the term began to be applied to all the nutrient matter. This terminology persists in referring to endospermic seeds as 'albuminous'. The nature of this material is used in both describing and classifying seeds, in addition to the embryo to endosperm size ratio. The endosperm may be considered to be farinaceous (or mealy) in which the cells are filled with starch, as for instance cereal grains, or not (non-farinaceous). The endosperm may also be referred to as 'fleshy' or 'cartilaginous' with thicker soft cells such as coconut, but may also be oily as in *Ricinus* (castor oil), *Croton* and Poppy. The endosperm is called 'horny' when the cell walls are thicker such as date and coffee, or 'ruminated' if mottled, as in nutmeg, palms and Annonaceae.

In most monocotyledons (such as grasses and palms) and some (endospermic or albuminous) dicotyledons (such as castor beans) the embryo is embedded in the endosperm (and nucellus, which the seedling will use upon germination. In the non-endospermic dicotyledons the endosperm is absorbed by the embryo as the latter grows within the developing seed, and the cotyledons of the embryo become filled with stored food. At maturity, seeds of these species have no endosperm and are also referred to as exalbuminous seeds. The exalbuminous seeds include the legumes (such as beans and peas), trees such as the oak and walnut, vegetables such as squash and radish, and sunflowers. According to Bewley and Black (1978), Brazil nut storage is in hypocotyl, this place of storage is uncommon among seeds. All gymnosperm seeds are albuminous.

Seed Coat

The seed coat develops from the maternal tissue, the integuments, originally surrounding the ovule. The seed coat in the mature seed can be a paper-thin layer (e.g. peanut) or something more substantial (e.g. thick and hard in honey locust and coconut), or fleshy as in the sarcotesta of pomegranate. The seed coat helps protect the embryo from mechanical injury, predators and drying out. Depending on its development, the seed coat is either bitegmic or unitegmic. Bitegmic seeds form a testa from the outer integument and a tegmen from the inner integument while unitegmic seeds have only one integument. Usually parts of the testa or tegmen form a hard protective mechanical layer. The mechanical layer may prevent water penetration and germination. Amongst the barriers may be the presence of lignified sclereids.

The outer integument has a number of layers, generally between four and eight organised into three layers: (a) outer epidermis, (b) outer pigmented zone of two to five layers containing tannin and starch, and (c) inner epidermis. The endotegmen is derived from the inner epidermis of the inner integument, the exotegmen from the outer surface of the inner integument. The endotesta is derived from the inner epidermis of the outer integument, and the outer layer of the testa from the outer surface of the outer integument is referred to as the exotesta. If the exotesta is also the mechanical layer, this is called an exotestal seed, but if the mechanical layer is the endotegmen, then the seed is endotestal. The exotesta may consist of one or more rows of cells that are elongated and pallisade like (e.g. Fabaceae), hence 'palisade exotesta'.

In addition to the three basic seed parts, some seeds have an appendage, an aril, a fleshy outgrowth of the funicle (funiculus), (as in yew and nutmeg) or an oily appendage, an elaiosome (as in *Corydalis*), or hairs (trichomes). In the latter example these hairs are the source of the textile crop cotton. Other seed appendages include the raphe (a ridge), wings, caruncles (a soft spongy outgrowth from the outer integument in the vicinity of the micropyle), spines, or tubercles.

A scar also may remain on the seed coat, called the hilum, where the seed was attached to the ovary wall by the funicle. Just below it is a small pore, representing the micropyle of the ovule.

Size and Seed Set

A collection of various vegetable and herb seeds

Seeds are very diverse in size. The dust-like orchid seeds are the smallest, with about one million seeds per gram; they are often embryonic seeds with immature embryos and no significant energy reserves. Orchids and a few other groups of plants are mycoheterotrophs which depend on mycorrhizal fungi for nutrition during germination and the early growth of the seedling. Some terrestrial orchid seedlings, in fact, spend the first few years of their lives deriving energy from the fungi and do not produce green leaves. At over 20 kg, the largest seed is the *coco de mer*. Plants that produce smaller seeds can generate many more seeds per flower, while plants with larger seeds invest more resources into those seeds and normally produce fewer seeds. Small seeds are quicker to ripen and can be dispersed sooner, so fall blooming plants often have small seeds. Many annual plants produce great quantities of smaller seeds; this helps to ensure at least a few will end in a favorable place for growth. Herbaceous perennials and woody plants often have larger seeds; they can produce seeds over many years, and larger seeds have more energy reserves for germination and seedling growth and produce larger, more established seedlings after germination.

Functions

Seeds serve several functions for the plants that produce them. Key among these functions are nourishment of the embryo, dispersal to a new location, and dormancy during unfavorable conditions. Seeds fundamentally are means of reproduction, and most seeds are the product of sexual reproduction which produces a remixing of genetic material and phenotype variability on which natural selection acts.

Embryo Nourishment

Seeds protect and nourish the embryo or young plant. They usually give a seedling a faster start than a sporeling from a spore, because of the larger food reserves in the seed and the multicellularity of the enclosed embryo.

Dispersal

Unlike animals, plants are limited in their ability to seek out favorable conditions for life and growth. As a result, plants have evolved many ways to disperse their offspring by dispersing their seeds. A seed must somehow "arrive" at a location and be there at a time favorable for germination and growth. When the fruits open and release their seeds in a regular way, it is called dehiscent, which is often distinctive for related groups of plants; these fruits include capsules, follicles, legumes, silicles and siliques. When fruits do not open and release their seeds in a regular fashion, they are called indehiscent, which include the fruits achenes, caryopsis, nuts, samaras, and utricles.

By Wind (Anemochory)

Dandelion seeds are contained within achenes, which can be carried long distances by the wind.

The seed pod of milkweed (*Asclepias syriaca*)

- Some seeds (e.g., pine) have a wing that aids in wind dispersal.
- The dustlike seeds of orchids are carried efficiently by the wind.

- Some seeds (e.g. milkweed, poplar) have hairs that aid in wind dispersal.

Other seeds are enclosed in fruit structures that aid wind dispersal in similar ways:

- Dandelion achenes have hairs.
- Maple samaras have two wings.

By Water (Hydrochory)

- Some plants, such as *Mucuna* and *Dioclea*, produce buoyant seeds termed sea-beans or drift seeds because they float in rivers to the oceans and wash up on beaches.

By Animals (Zoochory)

- Seeds (burrs) with barbs or hooks (e.g. acaena, burdock, dock) which attach to animal fur or feathers, and then drop off later.
- Seeds with a fleshy covering (e.g. apple, cherry, juniper) are eaten by animals (birds, mammals, reptiles, fish) which then disperse these seeds in their droppings.
- Seeds (nuts) are attractive long-term storable food resources for animals (e.g. acorns, hazelnut, walnut); the seeds are stored some distance from the parent plant, and some escape being eaten if the animal forgets them.

Myrmecochory is the dispersal of seeds by ants. Foraging ants disperse seeds which have appendages called elaiosomes (e.g. bloodroot, trilliums, acacias, and many species of Proteaceae). Elaiosomes are soft, fleshy structures that contain nutrients for animals that eat them. The ants carry such seeds back to their nest, where the elaiosomes are eaten. The remainder of the seed, which is hard and inedible to the ants, then germinates either within the nest or at a removal site where the seed has been discarded by the ants. This dispersal relationship is an example of mutualism, since the plants depend upon the ants to disperse seeds, while the ants depend upon the plants seeds for food. As a result, a drop in numbers of one partner can reduce success of the other. In South Africa, the Argentine ant (*Linepithema humile*) has invaded and displaced native species of ants. Unlike the native ant species, Argentine ants do not collect the seeds of *Mimetes cucullatus* or eat the elaiosomes. In areas where these ants have invaded, the numbers of *Mimetes* seedlings have dropped.

Dormancy

Seed dormancy has two main functions: the first is synchronizing germination with the optimal conditions for survival of the resulting seedling; the second is spreading germination of a batch of seeds over time so a catastrophe (e.g. late frosts, drought, herbivory) does not result in the death of all offspring of a plant (bet-hedging). Seed dormancy is defined as a seed failing to germinate under environmental conditions optimal for germination, normally when the environment is at a suitable temperature with proper soil moisture. This true dormancy or innate dormancy is therefore caused by conditions within the seed that prevent germination. Thus dormancy is a state of the seed, not of the environment. Induced dormancy, enforced dormancy or seed quiescence occurs when a seed fails to germinate because the external environmental conditions are inappropriate for germination, mostly in response to conditions being too dark or light, too cold or hot, or too dry.

Seed dormancy is not the same as seed persistence in the soil or on the plant, though even in scientific publications dormancy and persistence are often confused or used as synonyms.

Often, seed dormancy is divided into four major categories: exogenous; endogenous; combinational; and secondary. A more recent system distinguishes five classes: morphological, physiological, morphophysiological, physical and combinational dormancy.

Exogenous dormancy is caused by conditions outside the embryo, including:

- Physical dormancy or hard seed coats occurs when seeds are impermeable to water. At dormancy break, a specialized structure, the 'water gap', is disrupted in response to environmental cues, especially temperature, so water can enter the seed and germination can occur. Plant families where physical dormancy occurs include Anacardiaceae, Cannaceae, Convulvulaceae, Fabaceae and Malvaceae.
- Chemical dormancy considers species that lack physiological dormancy, but where a chemical prevents germination. This chemical can be leached out of the seed by rainwater or snow melt or be deactivated somehow. Leaching of chemical inhibitors from the seed by rain water is often cited as an important cause of dormancy release in seeds of desert plants, but little evidence exists to support this claim.

Endogenous dormancy is caused by conditions within the embryo itself, including:

- In morphological dormancy, germination is prevented due to morphological characteristics of the embryo. In some species, the embryo is just a mass of cells when seeds are dispersed; it is not differentiated. Before germination can take place, both differentiation and growth of the embryo have to occur. In other species, the embryo is differentiated but not fully grown (underdeveloped) at dispersal, and embryo growth up to a species specific length is required before germination can occur. Examples of plant families where morphological dormancy occurs are Apiaceae, Cycadaceae, Liliaceae, Magnoliaceae and Ranunculaceae.
- Morphophysiological dormancy includes seeds with underdeveloped embryos, and also have physiological components to dormancy. These seeds, therefore, require a dormancy-breaking treatments, as well as a period of time to develop fully grown embryos. Plant families where morphophysiological dormancy occurs include Apiaceae, Aquifoliaceae, Liliaceae, Magnoliaceae, Papaveraceae and Ranunculaceae. Some plants with morphophysiological dormancy, such as *Asarum* or *Trillium* species, have multiple types of dormancy, one affects radicle (root) growth, while the other affects plumule (shoot) growth. The terms "double dormancy" and "two-year seeds" are used for species whose seeds need two years to complete germination or at least two winters and one summer. Dormancy of the radicle (seedling root)is broken during the first winter after dispersal while dormancy of the shoot bud is broken during the second winter.
- Physiological dormancy means the embryo, due to physiological causes, cannot generate enough power to break through the seed coat, endosperm or other covering structures.

Dormancy is typically broken at cool wet, warm wet or warm dry conditions. Abscisic acid is usually the growth inhibitor in seeds, and its production can be affected by light.

  - Drying, in some plants, including a number of grasses and those from seasonally arid regions, is needed before they will germinate. The seeds are released, but need to have a lower moisture content before germination can begin. If the seeds remain moist after dispersal, germination can be delayed for many months or even years. Many herbaceous plants from temperate climate zones have physiological dormancy that disappears with drying of the seeds. Other species will germinate after dispersal only under very narrow temperature ranges, but as the seeds dry, they are able to germinate over a wider temperature range.

- In seeds with combinational dormancy, the seed or fruit coat is impermeable to water and the embryo has physiological dormancy. Depending on the species, physical dormancy can be broken before or after physiological dormancy is broken.

- Secondary dormancy* is caused by conditions after the seed has been dispersed and occurs in some seeds when nondormant seed is exposed to conditions that are not favorable to germination, very often high temperatures. The mechanisms of secondary dormancy are not yet fully understood, but might involve the loss of sensitivity in receptors in the plasma membrane.

The following types of seed dormancy do not involve seed dormancy, strictly speaking, as lack of germination is prevented by the environment, not by characteristics of the seed itself:

- Photodormancy or light sensitivity affects germination of some seeds. These photoblastic seeds need a period of darkness or light to germinate. In species with thin seed coats, light may be able to penetrate into the dormant embryo. The presence of light or the absence of light may trigger the germination process, inhibiting germination in some seeds buried too deeply or in others not buried in the soil.

- Thermodormancy is seed sensitivity to heat or cold. Some seeds, including cocklebur and amaranth, germinate only at high temperatures (30 °C or 86 °F); many plants that have seeds that germinate in early to midsummer have thermodormancy, so germinate only when the soil temperature is warm. Other seeds need cool soils to germinate, while others, such as celery, are inhibited when soil temperatures are too warm. Often, thermodormancy requirements disappear as the seed ages or dries.

Not all seeds undergo a period of dormancy. Seeds of some mangroves are viviparous; they begin to germinate while still attached to the parent. The large, heavy root allows the seed to penetrate into the ground when it falls. Many garden plant seeds will germinate readily as soon as they have water and are warm enough; though their wild ancestors may have had dormancy, these cultivated plants lack it. After many generations of selective pressure by plant breeders and gardeners, dormancy has been selected out.

For annuals, seeds are a way for the species to survive dry or cold seasons. Ephemeral plants are usually annuals that can go from seed to seed in as few as six weeks.

Persistence and Seed Banks

Germination

Germinating sunflower seedlings

Seed germination is a process by which a seed embryo develops into a seedling. It involves the reactivation of the metabolic pathways that lead to growth and the emergence of the radicle or seed root and plumule or shoot. The emergence of the seedling above the soil surface is the next phase of the plant's growth and is called seedling establishment.

Three fundamental conditions must exist before germination can occur. (1) The embryo must be alive, called seed viability. (2) Any dormancy requirements that prevent germination must be overcome. (3) The proper environmental conditions must exist for germination.

Seed viability is the ability of the embryo to germinate and is affected by a number of different conditions. Some plants do not produce seeds that have functional complete embryos, or the seed may have no embryo at all, often called empty seeds. Predators and pathogens can damage or kill the seed while it is still in the fruit or after it is dispersed. Environmental conditions like flooding or heat can kill the seed before or during germination. The age of the seed affects its health and germination ability: since the seed has a living embryo, over time cells die and cannot be replaced. Some seeds can live for a long time before germination, while others can only survive for a short period after dispersal before they die.

Seed vigor is a measure of the quality of seed, and involves the viability of the seed, the germination percentage, germination rate and the strength of the seedlings produced.

The germination percentage is simply the proportion of seeds that germinate from all seeds subject to the right conditions for growth. The germination rate is the length of time it takes for the seeds to germinate. Germination percentages and rates are affected by seed viability, dormancy and environmental effects that impact on the seed and seedling. In agriculture and horticulture quality seeds have high viability, measured by germination percentage plus the rate of germination. This is given as a percent of germination over a certain amount of time, 90% germination in 20 days, for example. 'Dormancy' is covered above; many plants produce seeds with varying degrees of dormancy, and different seeds from the same fruit can have different degrees of dormancy. It's possible to have seeds with no dormancy if they are dispersed right away and do not dry (if the seeds dry they go into physiological dormancy). There is great variation amongst plants and a dormant seed is still a viable seed even though the germination rate might be very low.

Environmental conditions affecting seed germination include; water, oxygen, temperature and light.

Three distinct phases of seed germination occur: water imbibition; lag phase; and radicle emergence.

In order for the seed coat to split, the embryo must imbibe (soak up water), which causes it to swell, splitting the seed coat. However, the nature of the seed coat determines how rapidly water can penetrate and subsequently initiate germination. The rate of imbibition is dependent on the permeability of the seed coat, amount of water in the environment and the area of contact the seed has to the source of water. For some seeds, imbibing too much water too quickly can kill the seed. For some seeds, once water is imbibed the germination process cannot be stopped, and drying then becomes fatal. Other seeds can imbibe and lose water a few times without causing ill effects, but drying can cause secondary dormancy.

Repair of DNA Damage

During seed dormancy, often associated with unpredictable and stressful environments, DNA damages accumulate as the seeds age. In rye seeds, the reduction of DNA integrity due to damage is associated with loss of seed viability during storage. Upon germination, seeds of *Vicia faba* undergo DNA repair. A plant DNA ligase that is involved in repair of single- and double-strand breaks during seed germination is an important determinant of seed longevity. Also, in Arabidopsis seeds, the activities of the DNA repair enzymes Poly ADP ribose polymerases (PARP) are likely needed for successful germination. Thus DNA damages that accumulate during dormancy appear to be a problem for seed survival, and the enzymatic repair of DNA damages during germination appears to be important for seed viability.

Inducing Germination

A number of different strategies are used by gardeners and horticulturists to break seed dormancy.

Scarification allows water and gases to penetrate into the seed; it includes methods to physically break the hard seed coats or soften them by chemicals, such as soaking in hot water or poking holes in the seed with a pin or rubbing them on sandpaper or cracking with a press or hammer. Sometimes fruits are harvested while the seeds are still immature and the seed coat is not fully developed and sown right away before the seed coat become impermeable. Under natural conditions, seed coats are worn down by rodents chewing on the seed, the seeds rubbing against rocks (seeds are moved by the wind or water currents), by undergoing freezing and thawing of surface water, or passing through an animal's digestive tract. In the latter case, the seed coat protects the seed from digestion, while often weakening the seed coat such that the embryo is ready to sprout when it is deposited, along with a bit of fecal matter that acts as fertilizer, far from the parent plant. Microorganisms are often effective in breaking down hard seed coats and are sometimes used by people as a treatment; the seeds are stored in a moist warm sandy medium for several months under nonsterile conditions.

Stratification, also called moist-chilling, breaks down physiological dormancy, and involves the addition of moisture to the seeds so they absorb water, and they are then subjected to a period of moist chilling to after-ripen the embryo. Sowing in late summer and fall and allowing to overwinter under cool conditions is an effective way to stratify seeds; some seeds respond more favorably to periods of oscillating temperatures which are a part of the natural environment.

Leaching or the soaking in water removes chemical inhibitors in some seeds that prevent germination. Rain and melting snow naturally accomplish this task. For seeds planted in gardens, running water is best—if soaked in a container, 12 to 24 hours of soaking is sufficient. Soaking longer, especially in stagnant water, can result in oxygen starvation and seed death. Seeds with hard seed coats can be soaked in hot water to break open the impermeable cell layers that prevent water intake.

Other methods used to assist in the germination of seeds that have dormancy include prechilling, predrying, daily alternation of temperature, light exposure, potassium nitrate, the use of plant growth regulators, such as gibberellins, cytokinins, ethylene, thiourea, sodium hypochlorite, and others. Some seeds germinate best after a fire. For some seeds, fire cracks hard seed coats, while in others, chemical dormancy is broken in reaction to the presence of smoke. Liquid smoke is often used by gardeners to assist in the germination of these species.

Sterile Seeds

Seeds may be sterile for few reasons: they may have been irradiated, unpollinated, cells lived past expectancy, or bred for the purpose.

Evolution and Origin of Seeds

The origin of seed plants is a problem that still remains unsolved. However, more and more data tends to place this origin in the middle Devonian. The description in 2004 of the proto-seed *Runcaria heinzelinii* in the Givetian of Belgium is an indication of that ancient origin of seed-plants. As with modern ferns, most land plants before this time reproduced by sending spores into the air, that would land and become whole new plants.

The first "true" seeds are described from the upper Devonian, which is probably the theater of their true first evolutionary radiation. The seed plants progressively became one of the major elements of nearly all ecosystems.

Economic Importance

Phaseolus vulgaris seeds are diverse in size, shape, and color.

Edible Seeds

Many seeds are edible and the majority of human calories comes from seeds, especially from cereals, legumes and nuts. Seeds also provide most cooking oils, many beverages and spices and some important food additives. In different seeds the seed embryo or the endosperm dominates and provides most of the nutrients. The storage proteins of the embryo and endosperm differ in their amino acid content and physical properties. For example, the gluten of wheat, important in providing the elastic property to bread dough is strictly an endosperm protein.

Seeds are used to propagate many crops such as cereals, legumes, forest trees, turfgrasses and pasture grasses. Particularly in developing countries, a major constraint faced is the inadequacy of the marketing channels to get the seed to poor farmers. Thus the use of farmer-retained seed remains quite common.

Seeds are also eaten by animals, and are fed to livestock. Many seeds are used as birdseed.

Poison and Food Safety

While some seeds are edible, others are harmful, poisonous or deadly. Plants and seeds often contain chemical compounds to discourage herbivores and seed predators. In some cases, these compounds simply taste bad (such as in mustard), but other compounds are toxic or break down into toxic compounds within the digestive system. Children, being smaller than adults, are more susceptible to poisoning by plants and seeds.

A deadly poison, ricin, comes from seeds of the castor bean. Reported lethal doses are anywhere from two to eight seeds, though only a few deaths have been reported when castor beans have been ingested by animals.

In addition, seeds containing amygdalin apple, apricot, bitter almond, peach, plum, cherry, quince, and others when consumed in sufficient amounts, may cause cyanide poisoning. Other seeds that contain poisons include annona, cotton, custard apple, datura, uncooked durian, golden chain, horse-chestnut, larkspur, locoweed, lychee, nectarine, rambutan, rosary pea, sour sop, sugar apple, wisteria, and yew. The seeds of the strychnine tree are also poisonous, containing the poison strychnine.

The seeds of many legumes, including the common bean (*Phaseolus vulgaris*), contain proteins called lectins which can cause gastric distress if the beans are eaten without cooking. The common bean and many others, including the soybean, also contain trypsin inhibitors which interfere with the action of the digestive enzyme trypsin. Normal cooking processes degrade lectins and trypsin inhibitors to harmless forms.

Other Uses

Cotton fiber grows attached to cotton plant seeds. Other seed fibers are from kapok and milkweed.

Many important nonfood oils are extracted from seeds. Linseed oil is used in paints. Oil from jojoba and crambe are similar to whale oil.

Seeds are the source of some medicines including castor oil, tea tree oil and the cancer drug, Laetrile.

Many seeds have been used as beads in necklaces and rosaries including Job's tears, Chinaberry, rosary pea, and castor bean. However, the latter three are also poisonous.

Other seed uses include:

- Seeds once used as weights for balances.
- Seeds used as toys by children, such as for the game Conkers.
- Resin from *Clusia rosea* seeds used to caulk boats.
- Nematicide from milkweed seeds.
- Cottonseed meal used as animal feed and fertilizer.

Seed Records

The massive fruit of the coco de mer

- The oldest viable carbon-14-dated seed that has grown into a plant was a Judean date palm seed about 2,000 years old, recovered from excavations at Herod the Great's palace on Masada in Israel. It was germinated in 2005. (A reported regeneration of *Silene stenophylla* (narrow-leafed campion) from material preserved for 31,800 years in the Siberian permafrost was achieved using fruit tissue, not seed.)
- The largest seed is produced by the coco de mer, or "double coconut palm", *Lodoicea maldivica*. The entire fruit may weigh up to 23 kilograms (50 pounds) and usually contains a single seed.
- The smallest seeds are produced by epiphytic orchids. They are only 85 micrometers long, and weigh 0.81 micrograms. They have no endosperm and contain underdeveloped embryos.
- The earliest fossil seeds are around 365 million years old from the Late Devonian of West Virginia. The seeds are preserved immature ovules of the plant *Elkinsia polymorpha*.

In Religion

The Book of Genesis in the Old Testament begins with an explanation of how all plant forms began:

And God said, Let the earth bring forth grass, the herb yielding seed, and the fruit tree yielding fruit after his kind, whose seed is in itself, upon the earth: and it was so. And the earth brought forth grass, and herb yielding seed after its kind, and the tree yielding fruit, whose seed was in itself, after its kind: and God saw that it was good. And the evening and the morning were the third day.

The Quran speaks about seed germination:

It is Allah Who causeth the seed-grain and the date-stone to split and sprout. He causeth the living to issue from the dead, and He is the one to cause the dead to issue from the living. That is Allah: then how are ye deluded away from the truth?

Seed Dispersal

Epilobium hirsutum seed head dispersing seeds

Seed dispersal is the movement or transport of seeds away from the parent plant. Plants have very limited mobility and consequently rely upon a variety of dispersal vectors to transport their propagules, including both abiotic and biotic vectors. Seeds can be dispersed away from the parent plant individually or collectively, as well as dispersed in both space and time. The patterns of seed dispersal are determined in large part by the dispersal mechanism and this has important implications for the demographic and genetic structure of plant populations, as well as migration patterns and species interactions. There are five main modes of seed dispersal: gravity, wind, ballistic, water, and by animals. Some plants are serotinous and only disperse their seeds in response to an environmental stimulus.

Benefits

Seed dispersal is likely to have several benefits for plant species. First, seed survival is often higher away from the parent plant. This higher survival may result from the actions of density-dependent seed and seedling predators and pathogens, which often target the high concentrations of seeds beneath adults. Competition with adult plants may also be lower when seeds are transported away from their parent.

Seed dispersal also allows plants to reach specific habitats that are favorable for survival, a hypothesis known as directed dispersal. For example, *Ocotea endresiana* (Lauraceae) is a tree species from Latin America which is dispersed by several species of birds, including the three-wattled bellbird. Male bellbirds perch on dead trees in order to attract mates, and often defecate seeds beneath these perches where the seeds have a high chance of survival because of high light conditions and escape from fungal pathogens. In the case of fleshy-fruited plants, seed-dispersal in animal guts (endozoochory) often enhances the amount, the speed, and the asynchrony of germination, which can have important plant benefits.

Seeds dispersed by ants (myrmecochory) are not only dispersed short distances but are also buried underground by the ants. These seeds can thus avoid adverse environmental effects such as fire or drought, reach nutrient-rich microsites and survive longer than other seeds. These features are peculiar to myrmecochory, which may thus provide additional benefits not present in other dispersal modes.

Finally, at another scale, seed dispersal may allow plants to colonize vacant habitats and even new geographic regions. Dispersal distances and deposition sites depend on the movement range of the disperser, and longer dispersal distances are sometimes accomplished through diplochory, the sequential dispersal by two or more different dispersal mechanisms. In fact, recent evidence suggests that the majority of seed dispersal events involves more than one dispersal phase.

Types

Seed dispersal is sometimes split into *autochory* (when dispersal is attained using the plant's own means) and *allochory* (when obtained through external means).

Autochory

Autochorous plants disperse their seed without any help from an external vector, as a result this limits plants considerably as to the distance they can disperse their seed. Two other types of autochory not described in detail here are blastochory, where the stem of the plant crawls along the ground to deposit its seed far from the base of the plant, and herpochory (the seed crawls by means of trichomes and changes in humidity).

Gravity

Barochory or the plant use of gravity for dispersal is a simple means of achieving seed dispersal. The effect of gravity on heavier fruits causes them to fall from the plant when ripe. Fruits exhibiting this type of dispersal include apples, coconuts and passionfruit and those with harder shells (which often roll away from the plant to gain more distance). Gravity dispersal also allows for later transmission by water or animal.

Ballistic Dispersal

Ballochory is a type of dispersal where the seed is forcefully ejected by explosive dehiscence of the fruit. Often the force that generates the explosion results from turgor pressure within the fruit or due to internal tensions within the fruit. Some examples of plants which disperse their seeds autochorously include: *Impatiens spp.*, *Arceuthobium spp.*, *Ecballium spp.*, *Geranium spp.*, *Cardamine hirsuta* and others. An exceptional example of ballochory is *Hura crepitans*—this plant is commonly called the dynamite tree due to the sound of the fruit exploding. The explosions are powerful enough to throw the seed up to 100 meters.

Allochory

Allochory refers to any of many types of seed dispersal where a vector or secondary agent is used to disperse seeds. This vectors may include wind, water, animals or others.

Wind

Wind dispersal of dandelion seeds

Entada phaseoloides – Hydrochory

Wind dispersal (*anemochory*) is one of the more primitive means of dispersal. Wind dispersal can take on one of two primary forms: seeds can float on the breeze or alternatively, they can flutter to the ground. The classic examples of these dispersal mechanisms, in the temperate northern hemisphere, include dandelions, which have a feathery pappus attached to their seeds and can be dispersed long distances, and maples, which have winged seeds (samaras) and flutter to the ground. An important constraint on wind dispersal is the need for abundant seed production to maximize the likelihood of a seed landing in a site suitable for germination. There are also strong evolutionary constraints on this dispersal mechanism. For instance, Cody and Overton (1996) found that species in the Asteraceae on islands tended to have reduced dispersal capabilities (i.e., larger seed mass and smaller pappus) relative to the same species on the mainland. Also, *Helonias bullata*, a species perennial herb native to the United States, evolved to utilize wind dispersal as the primary seed dispersal mechanism; however, limited wind in its habitat prevents the seeds to successfully disperse away from its parents, resulting in clusters of population. Reliance on wind dispersal is common among many weedy or ruderal species. Unusual mechanisms of wind dispersal include tumbleweeds, where the entire plant is blown by the wind. *Physalis* fruits, when not fully ripe, may sometimes be dispersed by wind due to the space between the fruit and the covering calyx which acts as air bladder.

Water

Many aquatic (water dwelling) and some terrestrial (land dwelling) species use *hydrochory*, or seed dispersal through water. Seeds can travel for extremely long distances, depending on the specific mode of fart water dispersal; this especially applies to fruits which are waterproof and float.

The water lily is an example of such a plant. Water lilies' flowers make a fruit that floats in the water for a while and then drops down to the bottom to take root on the floor of the pond. The seeds of palm trees can also be dispersed by water. If they grow near oceans, the seeds can be transported by ocean currents over long distances, allowing the seeds to be dispersed as far as other continents.

Mangrove trees grow directly out of the water; when their seeds are ripe they fall from the tree and grow roots as soon as they touch any kind of soil. During low tide, they might fall in soil instead of water and start growing right where they fell. If the water level is high, however, they can be carried far away from where they fell. Mangrove trees often make little islands as dirt and other things collect in their roots, making little bodies of land.

A special review for oceanic waters hydrochory can be seen at oceanic dispersal.

The "bill" and seed dispersal mechanism of *Geranium pratense*

By Animals

The small hooks on the surface of a bur enable attachment to animal fur for dispersion.

Animals can disperse plant seeds in several ways, all named *zoochory*. Seeds can be transported on the outside of vertebrate animals (mostly mammals), a process known as *epizoochory*. Plant species transported externally by animals can have a variety of adaptations for dispersal, including adhesive mucus, and a variety of hooks, spines and barbs. A typical example of an epizoochorous plant is *Trifolium angustifolium*, a species of Old World clover which adheres to animal fur by means of stiff hairs covering the seed. Epizoochorous plants tend to be herbaceous plants, with many representative species in the families Apiaceae and Asteraceae. However, epizoochory is a relatively rare dispersal syndrome for plants as a whole; the percentage of plant species with seeds adapted for transport on the outside of animals is estimated to be below 5%. Nevertheless, epizoochorous transport can be highly effective if seeds attach to wide-ranging animals. This form of seed dispersal has been implicated in rapid plant migration and the spread of invasive species.

Seed dispersal via ingestion by vertebrate animals (mostly birds and mammals), or *endozoochory*, is the dispersal mechanism for most tree species. Endozoochory is generally a coevolved mutualistic relationship in which a plant surrounds seeds with an edible, nutritious fruit as a good food for animals that consume it. Birds and mammals are the most important seed dispersers, but a wide variety of other animals, including turtles and fish, can transport viable seeds. The exact percentage of tree species dispersed by endozoochory varies between habitats, but can range to over 90% in some tropical rainforests. Seed dispersal by animals in tropical rainforests has received much attention, and this interaction is considered an important force shaping the ecology and evolution of vertebrate and tree populations. In the tropics, large animal seed dispersers (such as tapirs, chimpanzees and hornbills) may disperse large seeds with few other seed dispersal agents. The extinction of these large frugivores from poaching and habitat loss may have negative effects on the tree populations that depend on them for seed dispersal.

Seed dispersal by ants (*myrmecochory*) is a dispersal mechanism of many shrubs of the southern hemisphere or understorey herbs of the northern hemisphere. Seeds of myrmecochorous plants have a lipid-rich attachment called the elaiosome, which attracts ants. Ants carry such seeds into their colonies, feed the elaiosome to their larvae and discard the otherwise intact seed in an underground chamber. Myrmecochory is thus a coevolved mutualistic relationship between plants and seed-disperser ants. Myrmecochory has independently evolved at least 100 times in flowering plants and is estimated to be present in at least 11 000 species, but likely up to 23 000 or 9% of all species of flowering plants. Myrmecochorous plants are most frequent in the fynbos vegetation of the Cape Floristic Region of South Africa, the kwongan vegetation and other dry habitat types of Australia, dry forests and grasslands of the Mediterranean region and northern temperate forests of western Eurasia and eastern North America, where up to 30–40% of understorey herbs are myrmecochorous.

Seed predators, which include many rodents (such as squirrels) and some birds (such as jays) may also disperse seeds by hoarding the seeds in hidden caches. The seeds in caches are usually well-protected from other seed predators and if left uneaten will grow into new plants. In addition, rodents may also disperse seeds via seed spitting due to the presence of secondary metabolites in ripe fruits. Finally, seeds may be secondarily dispersed from seeds deposited by primary animal dispersers, a process known as diplochory. For example, dung beetles are known to disperse seeds from clumps of feces in the process of collecting dung to feed their larvae.

Other types of zoochory are *chiropterochory* (by bats), *malacochory* (by molluscs, mainly terrestrial snails), *ornithochory* (by birds) and *saurochory* (by non-bird sauropsids). Zoochory can occur in more than one phase, for example through *diploendozoochory*, where a primary disperser (an animal that ate a seed) along with the seeds it is carrying is eaten by a predator that then carries the seed further before depositing it.

By Humans

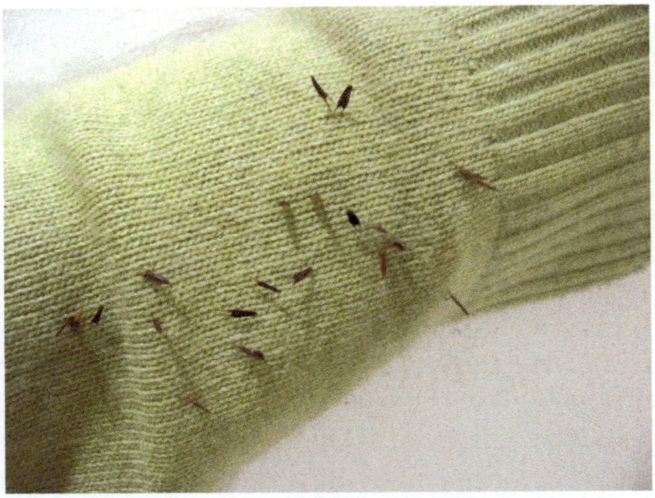

Epizoochory in *Bidens tripartita*; the seeds have attached to the clothes of a human.

Dispersal by humans (*anthropochory*) used to be seen as a form of dispersal by animals. Recent research points out that human dispersers differ from animal dispersers by a much higher mobility based on the technical means of human transport. Dispersal by humans on the one hand may act on large geographical scales and lead to invasive species. On the other hand, dispersal by humans also acts on smaller, regional scales and drives the dynamics of existing biological populations. Humans may disperse seeds by many various means and some surprisingly high distances have been repeatedly measured. Examples are: dispersal on human clothes (up to 250 m), on shoes (up to 5 km) or by cars (regularly ~ 250 m, singles cases > 100 km).

Deliberate seed dispersal also occurs as seed bombing. This has risks as unsuitable provenance may introduce genetically unsuitable plants to new environments.

Consequences

Seed dispersal has many consequences for the ecology and evolution of plants. Dispersal is necessary for species migrations, and in recent times dispersal ability is an important factor in whether or not a species transported to a new habitat by humans will become an invasive species. Dispersal is also predicted to play a major role in the origin and maintenance of species diversity. For example, myrmecochory increased the rate of diversification more than twofold in plant groups in which it has evolved because myrmecochorous lineages contain more than twice as many species as their non-myrmecochorous sister groups. Dispersal of seeds away from the parent organism has a central role in two major theories for how biodiversity is maintained in natural ecosystems, the Janzen-Connell hypothesis and recruitment limitation. Seed dispersal is essential in allowing forest migration of flowering plants.

In addition, the speed and direction of wind are highly influential in the dispersal process and in turn the deposition patterns of floating seeds in the stagnant water bodies. The transportation of seeds is led by the wind direction. This effects colonization situated on the banks of a river or to wetlands adjacent to streams relative to the distinct wind directions. The wind dispersal process can also effect connections between water bodies. Essentially, wind plays a larger role in the dispersal of waterborne seeds in a short period of time, days and seasons, but the ecological process allows the process to become balanced throughout a time period of several years. The time period of which the dispersal occurs is essential when considering the consequences of wind on the ecological process.

Seedling

A seedling is a young plant sporophyte developing out of a plant embryo from a seed. Seedling development starts with germination of the seed. A typical young seedling consists of three main parts: the radicle (embryonic root), the hypocotyl (embryonic shoot), and the cotyledons (seed leaves). The two classes of flowering plants (angiosperms) are distinguished by their numbers of seed leaves: monocotyledons (monocots) have one blade-shaped cotyledon, whereas dicotyledons (dicots) possess two round cotyledons. Gymnosperms are more varied. For example, pine seedlings have up to eight cotyledons. The seedlings of some flowering plants have no cotyledons at all. These are said to be acotyledons.

Grass seedlings, 150-minute time lapse

The plumule is the part of a seed embryo that develops into the shoot bearing the first true leaves of a plant. In most seeds, for example the sunflower, the plumule is a small conical structure without any leaf structure. Growth of the plumule does not occur until the cotyledons have grown above ground. This is epigeal germination. However, in seeds such as the broad bean, a leaf structure is visible on the plumule in the seed. These seeds develop by the plumule growing up through the soil with the cotyledons remaining below the surface. This is known as hypogeal germination.

Photomorphogenesis and Etiolation

Dicot seedlings grown in the light develop short hypocotyls and open cotyledons exposing the epicotyl. This is also referred to as photomorphogenesis. In contrast, seedlings grown in the dark develop long hypocotyls and their cotyledons remain closed around the epicotyl in an *apical hook*.

This is referred to as skotomorphogenesis or etiolation. Etiolated seedlings are yellowish in color as chlorophyll synthesis and chloroplast development depend on light. They will open their cotyledons and turn green when treated with light.

In a natural situation, seedling development starts with skotomorphogenesis while the seedling is growing through the soil and attempting to reach the light as fast as possible. During this phase, the cotyledons are tightly closed and form the *apical hook* to protect the shoot apical meristem from damage while pushing through the soil. In many plants, the seed coat still covers the cotyledons for extra protection.

Upon breaking the surface and reaching the light, the seedling's developmental program is switched to photomorphogenesis. The cotyledons open upon contact with light (splitting the seed coat open, if still present) and become green, forming the first photosynthetic organs of the young plant. Until this stage, the seedling lives off the energy reserves stored in the seed. The opening of the cotyledons exposes the shoot apical meristem and the *plumule* consisting of the first *true leaves* of the young plant.

The seedlings sense light through the light receptors phytochrome (red and far-red light) and cryptochrome (blue light). Mutations in these photo receptors and their signal transduction components lead to seedling development that is at odds with light conditions, for example seedlings that show photomorphogenesis when grown in the dark.

Seedling Growth and Maturation

Seedling of a dicot, *Nandina domestica*, showing two green cotyledon leaves, and the first "true" leaf with its distinct leaflets and red-green color.

Once the seedling starts to photosynthesize, it is no longer dependent on the seed's energy reserves. The apical meristems start growing and give rise to the root and shoot. The first "true" leaves expand and can often be distinguished from the round cotyledons through their species-dependent distinct shapes. While the plant is growing and developing additional leaves, the cotyledons eventually senesce and fall off. Seedling growth is also affected by mechanical stimulation, such as by wind or other forms of physical contact, through a process called thigmomorphogenesis.

Temperature and light intensity interact as they affect seedling growth; at low light levels about 40 lumens/m² a day/night temperature regime of 28 °C/13 °C is effective (Brix 1972). A photoperiod shorter than 14 hours causes growth to stop, whereas a photoperiod extended with low light intensities to 16 h or more brings about continuous (free) growth. Little is gained by using more than 16 h of low light intensity once seedlings are in the free growth mode. Long photoperiods using high light intensities from 10,000 to 20,000 lumens/m² increase dry matter production, and increasing the photoperiod from 15 to 24 hours may double dry matter growth (Pollard and Logan 1976, Carlson 1979).

The effects of carbon dioxide enrichment and nitrogen supply on the growth of white spruce and trembling aspen were investigated by Brown and Higginbotham (1986). Seedlings were grown in controlled environments with ambient or enriched atmospheric CO_2 (350 or 750 *f*1/L, respectively) and with nutrient solutions with high, medium, and low N content (15.5, 1.55, and 0.16 mM). Seedlings were harvested, weighed, and measured at intervals of less than 100 days. N supply strongly affected biomass accumulation, height, and leaf area of both species. In white spruce only, the root weight ratio (RWR) was significantly increased with the low-nitrogen regime. CO_2 enrichment for 100 days significantly increased the leaf and total biomass of white spruce seedlings in the high-N regime, RWR of seedlings in the medium-N regime, and root biomass of seedlings in the low-N regime.

First-year seedlings typically have high mortality rates, drought being the principal cause, with roots having been unable to develop enough to maintain contact with soil sufficiently moist to prevent the development of lethal seedling water stress. Somewhat paradoxically, however, Eis (1967a) observed that on both mineral and litter seedbeds, seedling mortality was greater in moist habitats (alluvium and *Aralia–Dryopteris*) than in dry habitats (*Cornus*–Moss). He commented that in dry habitats after the first growing season surviving seedlings appeared to have a much better chance of continued survival than those in moist or wet habitats, in which frost heave and competition from lesser vegetation became major factors in later years. The annual mortality documented by Eis (1967a) is instructive.

Pests and Diseases

Seedlings are particularly vulnerable to attack by pests and diseases and can consequently experience high mortality rates. Pests and diseases which are especially damaging to seedlings include damping off, cutworms, slugs and snails.

Transplanting

Seedlings are generally transplanted when the first pair of true leaves appear. A shade may be provided if the area is arid or hot. A commercially available vitamin hormone concentrate may be used to avoid transplant shock which may contain thiamine hydrochloride, naphthly acetic acid and indole butyric acid.

Cotyledon

A cotyledon is a significant part of the embryo within the seed of a plant, and is defined by the Oxford English Dictionary as "The primary leaf in the embryo of the higher plants (Phanerogams); the seed-leaf." Upon germination, the cotyledon may become the embryonic first leaves of a seedling. The number of cotyledons present is one characteristic used by botanists to classify the flowering plants (angiosperms). Species with one cotyledon are called monocotyledonous ("monocots"). Plants with two embryonic leaves are termed dicotyledonous ("dicots") and placed in the class Magnoliopsida.

Cotyledon from a Judas-tree (*Cercis siliquastrum*) seedling

In the case of dicot seedlings whose cotyledons are photosynthetic, the cotyledons are functionally similar to leaves. However, true leaves and cotyledons are developmentally distinct. Cotyledons are formed during embryogenesis, along with the root and shoot meristems, and are therefore present in the seed prior to germination. True leaves, however, are formed post-embryonically (i.e. after germination) from the shoot apical meristem, which is responsible for generating subsequent aerial portions of the plant.

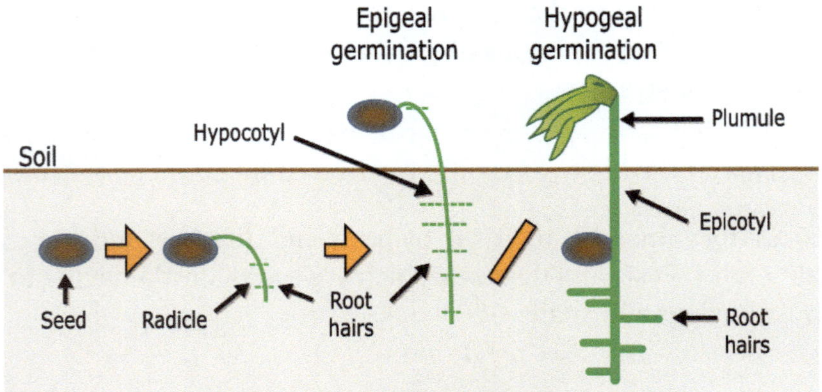

Schematic of epigeal vs hypogeal germination

The cotyledon of grasses and many other monocotyledons is a highly modified leaf composed of a *scutellum* and a *coleoptile*. The scutellum is a tissue within the seed that is specialized to absorb stored food from the adjacent endosperm. The coleoptile is a protective cap that covers the *plumule* (precursor to the stem and leaves of the plant).

Peanut seeds cut in half showing the embryos with cotyledons and primordial root.

Two-weeks-old cotyledons of Douglas-fir.

Gymnosperm seedlings also have cotyledons, and these are often variable in number (multicotyledonous), with from 2 to 24 cotyledons forming a whorl at the top of the hypocotyl (the embryonic stem) surrounding the plumule. Within each species, there is often still some variation in cotyledon numbers, e.g. Monterey pine (*Pinus radiata*) seedlings have 5–9, and Jeffrey pine (*Pinus jeffreyi*) 7–13 (Mirov 1967), but other species are more fixed, with e.g. Mediterranean cypress always having just two cotyledons. The highest number reported is for big-cone pinyon (*Pinus maximartinezii*), with 24 (Farjon & Styles 1997).

The cotyledons may be ephemeral, lasting only days after emergence, or persistent, enduring at least a year on the plant. The cotyledons contain (or in the case of gymnosperms and monocotyledons, have access to) the stored food reserves of the seed. As these reserves are used up, the cotyledons may turn green and begin photosynthesis, or may wither as the first true leaves take over food production for the seedling.

Epigeal Versus Hypogeal Development

Cotyledons may be either epigeal, expanding on the germination of the seed, throwing off the seed shell, rising above the ground, and perhaps becoming photosynthetic; or hypogeal, not expanding, remaining below ground and not becoming photosynthetic. The latter is typically the case where the cotyledons act as a storage organ, as in many nuts and acorns.

Hypogeal plants have (on average) significantly larger seeds than epigeal ones. They are also capable of surviving if the seedling is clipped off, as meristem buds remain underground (with epigeal plants, the meristem is clipped off if the seedling is grazed). The tradeoff is whether the plant should produce a large number of small seeds, or a smaller number of seeds which are more likely to survive.

Related plants show a mixture of hypogeal and epigeal development, even within the same plant family. Groups which contain both hypogeal and epigeal species include, for example, the Araucariaceae family of Southern Hemisphere conifers, the Fabaceae (pea family), and the genus *Lilium*. The frequently garden grown common bean - Phaseolus vulgaris - is epigeal while the closely related runner bean - Phaseolus coccineus - is hypogeal.

History

The term *cotyledon* was coined by Marcello Malpighi (1628–1694). John Ray was the first botanist to recognize that some plants have two and others only one, and eventually the first to recognize the immense importance of this fact to systematics, in *Methodus plantarum* (1682).

Theophrastus (3rd or 4th century BC) and Albertus Magnus (13th century) may also have recognized the distinction between the dicotyledons and monocotyledons.

Spermatophyte

The spermatophytes, also known as phanerogams or phenogamae, comprise those plants that produce seeds, hence the alternative name seed plants. They are a subset of the embryophytes or land plants. The term *phanerogams* or *phanerogamae* is derived from the Greek, *phanerós* meaning "visible", in contrast to the cryptogamae from Greek *kryptós* = "hidden" together with the suffix, *gameein*, "to marry". These terms distinguished those plants with hidden sexual organs (cryptogamae) from those with visible sexual organs (phanerogamae).

Description

The living spermatophytes form five groups, the first four of which were traditionally grouped as "gymnosperms":

- cycads, a subtropical and tropical group of plants with a large crown of compound leaves and a stout trunk,
- *Ginkgo*, a single living species of tree,
- conifers, cone-bearing trees and shrubs,
- gnetophytes, woody plants in the genera *Ephedra*, *Gnetum*, and *Welwitschia*

- angiosperms, (or magnoliophyta) the flowering plants, a large group including many familiar plants in a wide variety of habitats.

In addition to the taxa listed above, the fossil (old creature) record contains evidence of many extinct taxa of seed plants. The so-called "seed ferns" (Pteridospermae) were one of the earliest successful groups of land plants, and forests dominated by seed ferns were prevalent in the late Paleozoic. *Glossopteris* was the most prominent tree genus in the ancient southern supercontinent of Gondwana during the Permian period. By the Triassic period, seed ferns had declined in ecological importance, and representatives of modern gymnosperm groups were abundant and dominant through the end of the Cretaceous, when angiosperms radiated.

Evolution

A whole genome duplication event in the ancestor of seed plants occurred about 319 million years ago. This gave rise to a series of evolutionary changes that resulted in the origin of seed plants.

A middle Devonian precursor to seed plants from Belgium has been identified predating the earliest seed plants by about 20 million years. *Runcaria*, small and radially symmetrical, is an integumented megasporangium surrounded by a cupule. The megasporangium bears an unopened distal extension protruding above the mutlilobed integument. It is suspected that the extension was involved in anemophilous pollination. *Runcaria* sheds new light on the sequence of character acquisition leading to the seed. *Runcaria* has all of the qualities of seed plants except for a solid seed coat and a system to guide the pollen to the seed.

Seed Dormancy

A dormant seed is one that is unable to germinate in a specified period of time under a combination of environmental factors that are normally suitable for the germination of the non-dormant seed. Dormancy is a mechanism to prevent germination during unsuitable ecological conditions, when the probability of seedling survival is low.

One important function of most seeds is delayed germination, which allows time for dispersal and prevents germination of all the seeds at the same time. The staggering of germination safeguards some seeds and seedlings from suffering damage or death from short periods of bad weather or from transient herbivores; it also allows some seeds to germinate when competition from other plants for light and water might be less intense. Another form of delayed seed germination is seed quiescence, which is different from true seed dormancy and occurs when a seed fails to germinate because the external environmental conditions are too dry or warm or cold for germination. Many species of plants have seeds that delay germination for many months or years, and some seeds can remain in the soil seed bank for more than 50 years before germination. Some seeds have a very long viability period, and the oldest documented germinating seed was nearly 2000 years old based on radiocarbon dating.

Overview

True dormancy or innate dormancy is caused by conditions within the seed that prevent germination under normally ideal conditions. Often seed dormancy is divided into two major categories based on what part of the seed produces dormancy: exogenous and endogenous. There are three types of dormancy based on their mode of action: physical, physiological and morphological.

There have been a number of classification schemes developed to group different dormant seeds, but none have gained universal usage. Dormancy occurs because of a wide range of reasons that often overlap, producing conditions in which definitive categorization is not clear. Compounding this problem is that the same seed that is dormant for one reason at a given point may be dormant for another reason at a later point. Some seeds fluctuate from periods of dormancy to non dormancy, and despite the fact that a dormant seed appears to be static or inert, in reality they are still receiving and responding to environmental cues.

Exogenous Dormancy

Exogenous dormancy is caused by conditions outside the embryo and is often broken down into three subgroups:

Physical Dormancy

Dormancy that is caused by an impermeable seed coat is known as physical dormancy. Physical dormancy is the result of impermeable layer(s) that develops during maturation and drying of the seed or fruit. This impermeable layer prevents the seed from taking up water or gases. As a result, the seed is prevented from germinating until dormancy is broken. In natural systems, physical dormancy is broken by several factors including high temperatures, fluctuating temperatures, fire, freezing/thawing, drying or passage through the digestive tracts of animals. Physical dormancy is believed to have developed >100 mya.

Once physical dormancy is broken it cannot be reinstated i.e. the seed is unable to enter secondary dormancy following unfavourable conditions unlike seeds with physiological dormancy mechanisms. Therefore, the timing of the mechanisms that breaks physical dormancy is critical and must be tuned to environmental cues. This maximises the chances for germination occurring in conditions where the plant will successfully germinate, establish and eventually reproduce.

Physical dormancy has been identified in the seeds of plants across 15 angiosperm families including:

- Anacardiaceae
- Bixaceae
- Cannaceae (monocot)
- Cistaceae
- Cochlospermaceae

- Convolvulaceae
- Cucurbitaceae
- Dipterocarpaceae
- Geraniaceae
- Legumeinosae
- Malvaceae
- Nelumbonaceae
- Rhamnaceae
- Sarcolaenaceae
- Sapindaceae

Physical dormancy has not been recorded in any gymnosperms.

Generally, physical dormancy is the result of one or more palisade layers in the fruit or seed coat. These layers are lignified with malphigian cells tightly packed together and impregnated with water-repellent. In the Anacardiaceae and Nelumbonaceae families the seed coat is not well developed. Therefore, palisade layers in the fruit perform the functional role of preventing water uptake. While physical dormancy is a common feature, several species in these families do not have physical dormancy or produce non-dormant seeds.

Specialised structures, which function as a "water-gap", are associated with the impermeable layers of the seed to prevent the uptake of water. The water-gap is closed at seed maturity and is opened in response to the appropriate environmental signal. Breaking physical dormancy involves the disruption of these specialised structures within the seed, and acts as an environmental signal detector for germination. For example, legume (Fabaceae) seeds become permeable after the thin-walled cells of lens (water-gap structure). Following disrupted pulls apart to allow water entry into the seed.Other water gap structures include carpellary micropyle, bixoid chalazal plug, imbibition lid and the suberised 'stopper'.

In nature, the seed coats of physically dormant seeds are thought to become water permeable over time through repeated heating and cooling over many months-years in the soil seedbank. For example, the high and fluctuating temperatures during the dry season in northern Australia promote dormancy break in impermeable seeds of *Stylosanthes humilis* and *S.hamata* (Fabaceae).

Mechanical Dormancy

Mechanical dormancy occurs when seed coats or other coverings are too hard to allow the embryo to expand during germination. In the past this mechanism of dormancy was ascribed to a number of species that have been found to have endogenous factors for their dormancy instead. These endogenous factors include low embryo growth potential.

Chemical Dormancy

Includes growth regulators etc., that are present in the coverings around the embryo. They may be leached out of the tissues by washing or soaking the seed, or deactivated by other means. Other chemicals that prevent germination are washed out of the seeds by rainwater or snow melt.

Endogenous Dormancy

Endogenous dormancy is caused by conditions within the embryo itself, and it is also often broken down into three subgroups: physiological dormancy, morphological dormancy and combined dormancy, each of these groups may also have subgroups.

Physiological Dormancy

Physiological dormancy prevents embryo growth and seed germination until chemical changes occur. These chemicals include inhibitors that often retard embryo growth to the point where it is not strong enough to break through the seed coat or other tissues. Physiological dormancy is indicated when an increase in germination rate occurs after an application of gibberellic acid (GA3) or after Dry after-ripening or dry storage. It is also indicated when dormant seed embryos are excised and produce healthy seedlings: or when up to 3 months of cold (0–10 °C) or warm (=15 °C) stratification increases germination: or when dry after-ripening shortens the cold stratification period required. In some seeds physiological dormancy is indicated when scarification increases germination.

Physiological dormancy is broken when inhibiting chemicals are broken down or are no longer produced by the seed; often by a period of cool moist conditions, normally below (+4C) 39F, or in the case of many species in *Ranunculaceae* and a few others,(−5C) 24F. Abscisic acid is usually the growth inhibitor in seeds and its production can be affected by light. Some plants like Peony species have multiple types of physiological dormancy, one affects radicle (root) growth while the other affects plumule (shoot) growth. Seeds with physiological dormancy most often do not germinate even after the seed coat or other structures that interfere with embryo growth are removed. Conditions that affect physiological dormancy of seeds include:

- Drying; some plants including a number of grasses and those from seasonally arid regions need a period of drying before they will germinate, the seeds are released but need to have a lower moisture content before germination can begin. If the seeds remain moist after dispersal, germination can be delayed for many months or even years. Many herbaceous plants from temperate climate zones have physiological dormancy that disappears with drying of the seeds. Other species will germinate after dispersal only under very narrow temperature ranges, but as the seeds dry they are able to germinate over a wider temperature range.
- Photodormancy or light sensitivity affects germination of some seeds. These photoblastic seeds need a period of darkness or light to germinate. In species with thin seed coats, light may be able to penetrate into the dormant embryo. The presence of light or the absence of light may trigger the germination process, inhibiting germination in some seeds buried too deeply or in others not buried in the soil.

- Thermodormancy is seed sensitivity to heat or cold. Some seeds including cocklebur and amaranth germinate only at high temperatures (30C or 86F). Many plants that have seeds that germinate in early to mid summer have thermodormancy and germinate only when the soil temperature is warm. Other seeds need cool soils to germinate, while others like celery are inhibited when soil temperatures are too warm. Often thermodormancy requirements disappear as the seed ages or dries.

Seeds are classified as having deep physiological dormancy under these conditions: applications of GA3 does not increase germination; or when excised embryos produce abnormal seedlings; or when seeds require more than 3 months of cold stratification to germinate.

Morphological Dormancy

In morphological dormancy, the embryo is underdeveloped or undifferentiated. Some seeds have fully differentiated embryos that need to grow more before seed germination, or the embryos are not differentiated into different tissues at the time of fruit ripening.

- Immature embryos – some plants release their seeds before the tissues of the embryos have fully differentiated, and the seeds ripen after they take in water while on the ground, germination can be delayed from a few weeks to a few months.

Combined Dormancy

Seeds have both morphological and physiological dormancy.

- Morpho-physiological or morphophysiological dormancy occurs when seeds with underdeveloped embryos, also have physiological components to dormancy. These seeds therefore require dormancy-breaking treatments as well as a period of time to develop fully grown embryos.
- Intermediate simple
- Deep simple
- Deep simple epicotyl
- Deep simple double
- Intermediate complex
- Deep complex

Combinational Dormancy

Combinational dormancy occurs in some seeds, where dormancy is caused by both exogenous (physical) and endogenous (physiological) conditions. some *Iris* species have both hard impermeable seeds coats and physiological dormancy.

Secondary Dormancy

Secondary dormancy occurs in some non-dormant and post dormant seeds that are exposed to conditions that are not favorable for germination, like high temperatures. It is caused by conditions that occur after the seed has been dispersed. The mechanisms of secondary dormancy are not yet fully understood but might involve the loss of sensitivity in receptors in the plasma membrane.

Not all seeds undergo a period of dormancy, many species of plants release their seeds late in the year when the soil temperature is too low for germination or when the environment is dry. If these seeds are collected and sown in an environment that is warm enough, and/or moist enough, they will germinate. Under natural conditions non dormant seeds released late in the growing season wait until spring when the soil temperature rises or in the case of seeds dispersed during dry periods until it rains and there is enough soil moisture.

Seeds that do not germinate because they have fleshy fruits that retard germination are quiescent, not dormant.

Many garden plants have seeds that will germinate readily as soon as they have water and are warm enough, though their wild ancestors had dormancy. These cultivated plants lack seed dormancy because of generations of selective pressure by plant breeders and gardeners that grew and kept plants that lacked dormancy.

Seeds of some mangroves are viviparous and begin to germinate while still attached to the parent; they produce a large, heavy root, which allows the seed to penetrate into the ground when it falls.

Germination

Germination rate testing on the germination table

Germination is the process by which an organism grows from a seed or similar structure. The most common example of germination is the sprouting of a seedling from a seed of an angiosperm or gymnosperm. In addition, the growth of a sporeling from a spore, such as the spores of hyphae from fungal spores, is also germination. Thus, in a general sense, germination can be thought of as anything expanding into greater being from a small existence or germ.

Introduction

Germination is usually the growth of a plant contained within a seed; it results in the formation of the seedling, it is also the process of reactivation of metabolic machinery of the seed resulting in the emergence of radicle and plumule. The seed of a vascular plant is a small package produced in a fruit or cone after the union of male and female reproductive cells. All fully developed seeds contain an embryo and, in most plant species some store of food reserves, wrapped in a seed coat. Some plants produce varying numbers of seeds that lack embryos; these are called empty seeds and never germinate. Dormant seeds are ripe seeds that do not germinate because they are subject to external environmental conditions that prevent the initiation of metabolic processes and cell growth. Under proper conditions, the seed begins to germinate and the embryonic tissues resume growth, developing towards a seedling.

A seed tray used in horticulture for sowing and taking plant cuttings and growing plugs

Brassica campestris germinating seeds

Seed germination depends on both internal and external conditions. The most important external factors include right temperature, water, oxygen or air and sometimes light or darkness. Various plants require different variables for successful seed germination. Often this depends on the individual seed variety and is closely linked to the ecological conditions of a plant's natural habitat. For some seeds, their future germination response is affected by environmental conditions during seed formation; most often these responses are types of seed dormancy.

- Water is required for germination. Mature seeds are often extremely dry and need to take in significant amounts of water, relative to the dry weight of the seed, before cellular metabolism and growth can resume. Most seeds need enough water to moisten the seeds but not enough to soak them. The uptake of water by seeds is called imbibition, which leads to the swelling and the breaking of the seed coat. When seeds are formed, most plants store a food reserve with the seed, such as starch, proteins, or oils. This food reserve provides nourishment to the growing embryo. When the seed imbibes water, hydrolytic enzymes are activated which break down these stored food resources into metabolically useful chemicals. After the seedling emerges from the seed coat and starts growing roots and leaves, the seedling's food reserves are typically exhausted; at this point photosynthesis provides the energy needed for continued growth and the seedling now requires a continuous supply of water, nutrients, and light.

- Oxygen is required by the germinating seed for metabolism. Oxygen is used in aerobic respiration, the main source of the seedling's energy until it grows leaves. Oxygen is an atmospheric gas that is found in soil pore spaces; if a seed is buried too deeply within the soil or the soil is waterlogged, the seed can be oxygen starved. Some seeds have impermeable seed coats that prevent oxygen from entering the seed, causing a type of physical dormancy which is broken when the seed coat is worn away enough to allow gas exchange and water uptake from the environment.

- Temperature affects cellular metabolic and growth rates. Seeds from different species and even seeds from the same plant germinate over a wide range of temperatures. Seeds often have a temperature range within which they will germinate, and they will not do so above or below this range. Many seeds germinate at temperatures slightly above 60-75 F (16-24 C) [room-temperature if you live in a centrally heated house], while others germinate just above freezing and others germinate only in response to alternations in temperature between warm and cool. Some seeds germinate when the soil is cool 28-40 F (-2 - 4 C), and some when the soil is warm 76-90 F (24-32 C). Some seeds require exposure to cold temperatures (vernalization) to break dormancy. Some seeds in a dormant state will not germinate even if conditions are favorable. Seeds that are dependent on temperature to end dormancy have a type of physiological dormancy. For example, seeds requiring the cold of winter are inhibited from germinating until they take in water in the fall and experience cooler temperatures. Cold stratification is a process that induces the dormancy breaking prior to light emission that promotes germination . Four degrees Celsius is cool enough to end dormancy for most cool dormant seeds, but some groups, especially within the family Ranunculaceae and others, need conditions cooler than -5 C. Some seeds will only germinate after hot temperatures during a forest fire which cracks their seed coats; this is a type of physical dormancy.

Most common annual vegetables have optimal germination temperatures between 75-90 F (24-32 C), though many species (e.g. radishes or spinach) can germinate at significantly lower temperatures, as low as 40 F (4 C), thus allowing them to be grown from seeds in cooler climates. Suboptimal temperatures lead to lower success rates and longer germination periods.

- Light or darkness can be an environmental trigger for germination and is a type of physiological dormancy. Most seeds are not affected by light or darkness, but many seeds, including species found in forest settings, will not germinate until an opening in the canopy allows sufficient light for growth of the seedling.

Scarification mimics natural processes that weaken the seed coat before germination. In nature, some seeds require particular conditions to germinate, such as the heat of a fire (e.g., many Australian native plants), or soaking in a body of water for a long period of time. Others need to be passed through an animal's digestive tract to weaken the seed coat enough to allow the seedling to emerge.

Malted (germinated) barley grains

Dormancy

Some live seeds are dormant and need more time, and/or need to be subjected to specific environmental conditions before they will germinate. Seed dormancy can originate in different parts of the seed, for example, within the embryo; in other cases the seed coat is involved. Dormancy breaking often involves changes in membranes, initiated by dormancy-breaking signals. This generally occurs only within hydrated seeds. Factors affecting seed dormancy include the presence of certain plant hormones, notably abscisic acid, which inhibits germination, and gibberellin, which ends seed dormancy. In brewing, barley seeds are treated with gibberellin to ensure uniform seed germination for the production of barley malt.

Seedling Establishment

In some definitions, the appearance of the radicle marks the end of germination and the beginning of "establishment", a period that utilizes the food reserves stored in the seed. Germination and establishment as an independent organism are critical phases in the life of a plant when they are the most vulnerable to injury, disease, and water stress. The germination index can be used as an indicator of phytotoxicity in soils. The mortality between dispersal of seeds and completion of establishment can be so high that many species have adapted to produce huge numbers of seeds

Germination Rate and Germination Capacity

Germination of seedlings raised from seeds of eucalyptus after 3 days of sowing.

In agriculture and gardening, the germination rate describes how many seeds of a particular plant species, variety or seedlot are likely to germinate over a given period. It is a measure of germination time course and is usually expressed as a percentage, e.g., an 85% germination rate indicates that about 85 out of 100 seeds will probably germinate under proper conditions over the germination period given. The germination rate is useful for calculating the seed requirements for a given area or desired number of plants. In seed physiologists and seed scientists "germination rate" is the reciprocal of time taken for the process of germination to complete starting from time of sowing. On the other hand, the number of seed able to complete germination in a population (i.e. seed lot) is referred as germination capacity.

Dicot Germination

The part of the plant that first emerges from the seed is the embryonic root, termed the radicle or primary root. It allows the seedling to become anchored in the ground and start absorbing water. After the root absorbs water, an embryonic shoot emerges from the seed. This shoot comprises three main parts: the cotyledons (seed leaves), the section of shoot below the cotyledons (hypocotyl), and the section of shoot above the cotyledons (epicotyl). The way the shoot emerges differs among plant groups.

Epigeal

In epigeal germination (or epigeous germination), the *hypocotyl* elongates and forms a hook, pulling rather than pushing the cotyledons and apical meristem through the soil. Once it reaches the surface, it straightens and pulls the cotyledons and shoot tip of the growing seedlings into the air. Beans, tamarind and papaya are examples of plants that germinate this way.

Hypogeal

Germination can also be done by hypogeal germination (or hypogeous germination), where the epicotyl elongates and forms the hook. In this type of germination, the cotyledons stay underground where they eventually decompose. Peas, gram and mango, for example, germinate this way.

Monocot Germination

In monocot seeds, the embryo's radicle and cotyledon are covered by a coleorhiza and coleoptile, respectively. The coleorhiza is the first part to grow out of the seed, followed by the radicle. The coleoptile is then pushed up through the ground until it reaches the surface. There, it stops elongating and the first leaves emerge.

Precocious Germination

When a seed germinates without undergoing all four stages of seed development, i.e., globular, heart shape, torpedo shape, and cotyledonary stage, it is known as precocious germination.

Pollen Germination

Another germination event during the life cycle of gymnosperms and flowering plants is the germination of a pollen grain after pollination. Like seeds, pollen grains are severely dehydrated before being released to facilitate their dispersal from one plant to another. They consist of a protective coat containing several cells (up to 8 in gymnosperms, 2-3 in flowering plants). One of these cells is a tube cell. Once the pollen grain lands on the stigma of a receptive flower (or a female cone in gymnosperms), it takes up water and germinates. Pollen germination is facilitated by hydration on the stigma, as well as by the structure and physiology of the stigma and style. Pollen can also be induced to germinate *in vitro* (in a petri dish or test tube).

During germination, the tube cell elongates into a pollen tube. In the flower, the pollen tube then grows towards the ovule where it discharges the sperm produced in the pollen grain for fertilization. The germinated pollen grain with its two sperm cells is the mature male microgametophyte of these plants.

Self-incompatibility

Since most plants carry both male and female reproductive organs in their flowers, there is a high risk of self-pollination and thus inbreeding. Some plants use the control of pollen germination as a way to prevent this self-pollination. Germination and growth of the pollen tube involve molecular signaling between stigma and pollen. In self-incompatibility in plants, the stigma of certain plants can molecularly recognize pollen from the same plant and prevent it from germinating.

Spore Germination

Germination can also refer to the emergence of cells from resting spores and the growth of sporeling hyphae or thalli from spores in fungi, algae and some plants.

Conidia are asexual reproductive (reproduction without the fusing of gametes) spores of fungi which germinate under specific conditions. A variety of cells can be formed from the germinating conidia. The most common are germ tubes which grow and develop into hyphae. Another type of cell is a conidial anastomosis tube (CAT); these differ from germ tubes in that they are thinner, shorter, lack branches, exhibit determinate growth and home toward each other. Each cell is of a tubular shape, but the conidial anastomosis tube forms a bridge that allows fusion between conidia.

Resting Spores

In resting spores, germination that involves cracking the thick cell wall of the dormant spore. For example, in zygomycetes the thick-walled zygosporangium cracks open and the zygospore inside gives rise to the emerging sporangiophore. In slime molds, germination refers to the emergence of amoeboid cells from the hardened spore. After cracking the spore coat, further development involves cell division, but not necessarily the development of a multicellular organism (for example in the free-living amoebas of slime molds).

Ferns and Mosses

In plants such as bryophytes, ferns, and a few others, spores germinate into independent gametophytes. In the bryophytes (e.g., mosses and liverworts), spores germinate into protonemata, similar to fungal hyphae, from which the gametophyte grows. In ferns, the gametophytes are small, heart-shaped prothalli that can often be found underneath a spore-shedding adult plant.[

Epigeal Germination

Epigeal germination is a botanical term indicating that the germination of a plant takes place above the ground. An example of a plant with epigeal germination is the common bean (*Phaseolus vulgaris*). The opposite of epigeal is hypogeal (underground germination).

Germination

Epigeal germination implies that the cotyledons are pushed above ground. The hypocotyl (part of the stem below the cotyledon) elongates while the epicotyl (part of the stem above the cotyledon) stays the same length. In this way, the hypocotyl pushes the cotyledon upward.

Normally, the cotyledon itself contains very little nutrients in plants that show this kind of germination. Instead, the first leaflets are already folded up inside it, and photosynthesis starts to take place in it rather quickly.

Because the cotyledon is positioned above the ground it is much more vulnerable to damage like night-frost or grazing. The evolutionary strategy is that the plant produces a large number of seeds, of which statistically a number survive.

Plants that show epigeal germination need external nutrients rather quickly in order to develop, so they are more frequent on nutrient-rich soils. The plants also need relatively much sunlight for photosynthesis to take place. Therefore they can be found more often in the field, at the border of forests, or as pioneer species.

Plants that show epigeal germination grow relatively fast, especially in the first phase when the leaflets unfold. Because of this, they occur frequently in areas that experience regular flooding, for example at the river borders in the Amazon region. The fast germination enables the plant to develop before the next flooding takes place. After the faster first phase, the plant develops more slowly than plants that show hypogeal germination.

It is possible that within the same genus one species shows epigeal germination while another species shows hypogeal germination. Some genera in which this happens are:

- *Phaseolus*: the common bean (*Phaseolus vulgaris*) shows epigeal germination, whereas the runner bean (*Phaseolus coccineus*) shows hypogeal germination
- *Lilium*
- *Araucaria*: species in the section *Eutacta* show epigeal germination, whereas species in the section *Araucaria* show hypogeal germination

Phanerocotylar vs. Cryptocotylar

In 1965, botanist James A. Duke introduced the terms phanerocotylar and cryptocotylar as synonyms for epigeal and hypogeal respectively, because he didn't consider these terms etymologically correct. Later, it was discovered that there are rare cases of species where the germination is epigeal and cryptocotylar. Therefore, divisions have been proposed that take both factors into account.

Gametophyte

A gametophyte is a stage in the life cycle of plants and algae that undergo alternation of generations. It is a haploid multicellular organism that develops from a haploid spore that has one set of chromosomes. The gametophyte is the sexual phase in the life cycle of plants and algae. It develops sex organs that produce gametes, haploid sex cells that participate in fertilization to form a diploid zygote in which each cell has two sets of chromosomes. Cell division of the zygote results in a new diploid multicellular organism, the second stage in the life cycle known as the sporophyte, the function of which is to produce haploid spores by meiosis.

Several gametophytes growing in a terrarium.

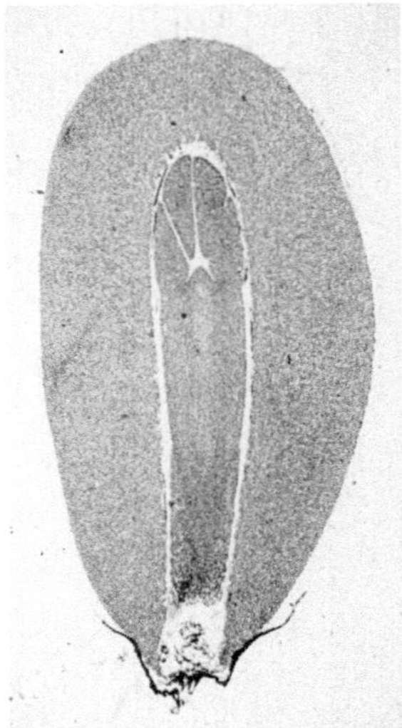

Pine gametophyte (outside) surrounding the embryo (inside)

Algae

In some multicellular green algae (*Ulva lactuca* is one example), red algae and brown algae sporophytes and gametophytes may be externally indistinguishable (isomorphic). In *Ulva* the gametes are isogamous, all of one size, shape and general morphology.

Land plants

In land plants, anisogamy is universal. As in animals, female and male gametes are called, respectively, *eggs* and *sperm*. Either the sporophyte or the gametophyte may be reduced (heteromorphic).]

Bryophytes

In bryophytes (mosses, liverworts, and hornworts), the gametophyte is the most visible stage of the life cycle. The bryophyte gametophyte is longer lived, nutritionally independent, and the sporophytes are typically attached to the gametophytes and dependent on them. When a moss spore germinates it grows to produce a filament of cells (called the protonema). The mature gametophyte of mosses develops into leafy shoots that produce sex organs (gametangia) that produce gametes. Eggs develop in archegonia and sperm in antheridia.

In some bryophyte groups such as many liverworts of the order Marchantiales, the gametes are produced on specialized structures called gametophores (or gametangiophores).

Ferns

In most ferns, gametophyte is a photosynthetic free living organism called a prothallus. For example, in the leptosporangiate fern *Dryopteris*, the gametophyte is a free-living, autotrophic prothallus that produces gametes and maintains the sporophyte during its early multicellular development. However, in some groups, notably the clade that includes Ophioglossaceae and *Psilotaceae*, the gametophytes are subterranean and subsist by forming mycotrophic relationships with fungi.

Lycophytes

Extant lycophytes produce several different types of gametophytes. In the families Lycopodiaceae and Huperziaceae, gametophytes are subterranean and mycotrophic, deriving nutrients from symbiosis with fungi. *Isoetes* and *Selaginella*, which are heterosporous, megagametophytes develop inside the megaspores, which crack open at the trilete suture to allow the male gametes to access the egg cells in the archegonia inside. The gametophytes of *Isoetes* appear to be similar in this respect to those of the extinct Carboniferous giant arborescent clubmosses, *Lepidodendron* and *Lepidostrobus*.

Seed Plants

By contrast, in seed plants (gymnosperms and angiosperms), gametophytes develop into multicellular organisms while still enclosed within the sporangium.

Vascular plants that produce only one type of spore are said to be homosporous. They have exosporic gametophytes—that is, the gametophyte is free-living and develops outside of the spore wall. Exosporic gametophytes are normally bisexual, capable of producing both sperm and eggs. In heterosporous vascular plants (plants that produce both microspores and megaspores), the gametophyte develops endosporically, within the spore wall. These gametophytes are unisexual, producing either sperm or eggs but not both.

All vascular plants are sporophyte dominant, and a trend toward smaller and more sporophyte-dependent female gametophytes is evident as land plants evolved towards reproduction by seeds.

Heteromorphy

In plants with heteromorphic gametophytes, there are two distinct kinds of gametophytes. Because the two gametophytes differ in form and function, they are termed *heteromorphic*, from *hetero-* "different" and *morph* "form". The egg producing gametophyte is known as a megagametophyte, because it is typically larger, and the sperm producing gametophyte is known as a microgametophyte. Gametophytes which produce egg and sperm on separate plants are termed Dioicous.

In heterosporous plants (water ferns, some lycophytes, as well as all gymnosperms and angiosperms), there are two distinct sporangia, each of which produces a single kind of spore and single kind of gametophyte. However, not all heteromorphic gametophytes come from heterosporous plants. That is, some plants have distinct egg-producing and sperm-producing gametophytes, but these gametophytes develop from the same kind of spore inside the same sporangium; *Sphaerocarpos* is an example of such a plant.

In the seed plants, the microgametophyte is called pollen. Seed plant microgametophytes consists of two or three cells when the pollen grains exit the sporangium. The megagametophyte develops within the megaspore of extant seedless vascular plants and within the megasporangium in a cone or flower in seed plants. In seed plants, the microgametophyte (pollen grain) travels to the vicinity of the egg cell (carried by a physical or animal vector), and produces two sperm by mitosis.

In gymnosperms the megagametophyte consists of several thousand cells and produces one to several archegonia, each with a single egg cell. The gametophyte becomes a food storage tissue in the seed.

In angiosperms, the megagametophyte is reduced to only a few nuclei and cells, and is sometimes called the embryo sac. A typical embryo sac contains seven cells and eight nuclei, one of which is the egg cell. Two nuclei fuse with a sperm nucleus to form endosperm, which becomes the food storage tissue in the seed.

Sprouting

Mixed bean sprouts

Sprouting is the practice of germinating seeds to be eaten raw or cooked. Sprouts can be germinated at home or produced industrially. They are a prominent ingredient of the raw food diet and common in Eastern Asian cuisine.

Soybean sprouts on a plate

Sprouting, like cooking, reduces anti-nutritional compounds in raw legumes. Raw lentils, for example, contain lectins, anti-nutritional proteins which can be reduced by sprouting or cooking. Sprouting is also applied on a large scale to barley as a part of the malting process. A downside to consuming raw sprouts is that the process of germinating seeds can also be conducive to harmful bacterial growth.

Seeds Suitable for Sprouting

All viable seeds can be sprouted, but some sprouts should not be eaten raw. The most common food sprouts include:

- Pulses (legumes; pea family):

  alfalfa, clover, fenugreek, lentil, pea, chickpea, mung bean and soybean (bean sprouts)

- Cereals:

  oat, wheat, maize (corn), rice, barley, and rye

- Pseudocereals:

  quinoa, amaranth and buckwheat

- Oilseeds:

  sesame, sunflower, almond, hazelnut, hemp, linseed, peanut

- Brassica (cabbage family)

  broccoli, cabbage, watercress, mustard, mizuna, radish, and daikon (*kaiware* sprouts), rocket (arugula), tatsoi, turnip

- Umbelliferous vegetables (parsley family) - these may be used more as microgreens than sprouts

  carrot, celery, fennel, parsley

- Allium (onions) - cannot really distinguish between microgreens.

  onion, leek, green onion (*me-negi* in Japanese cuisine)

- Other vegetables and herbs:

  spinach, lettuce, milk thistle, lemon grass.

Although whole oats can be sprouted, oat groats sold in food stores, which are dehulled and require steaming or roasting to prevent rancidity, will not sprout. Whole oats may have an indigestible hull which makes them difficult or even unfit for human consumption.

In the case of rice, the husk of the paddy is removed before sprouting. Brown rice is widely used for germination (GBR - Germinated Brown Rice) in Japan and other countries.

All the sprouts of the solanaceae (tomato, potato, paprika, aubergine or eggplant) and rhubarb cannot be eaten as sprouts, either cooked or raw, as they can be poisonous. Some sprouts can be cooked to remove the toxin, while others cannot.

With all seeds, care should be taken that they are intended for sprouting or human consumption rather than sowing. Seeds intended for sowing may be treated with chemical dressings. Several countries, such as New Zealand, also require that some varieties of imported edible seed be heat-treated, thus making them impossible to sprout. Quinoa in its natural state is very easy to sprout but when polished, or pre-cleaned of its saponin coating (becoming whiter), loses its power to germinate.

The Germination Process

Sprouting mung beans in a glass sprouter jar with a plastic sieve-lid

The germination process takes a few days and can be done at home manually, as a semi-automated process, or industrially on a large scale for commercial use.

Typically the seeds are first rinsed to remove soil and dirt and the mucilaginous substances produced by some seeds when they come in contact with water. Then they are soaked for 20 minutes to 12 hours, depending on the type and size of seed. The soaking increases the water content in the seeds and brings them out of quiescence. After draining and then rinsing seeds at regular intervals they germinate, or sprout.

For home sprouting, the seeds are soaked (big seeds) or moistened (small), then left at room temperature (13 to 21 °C or 55 to 70 °F) in a sprouting vessel. Many different types of vessels can be used. One type is a simple glass jar with a piece of cloth or nylon window screen secured over its rim. "Tiered" clear plastic sprouters are commercially available, allowing a number of "crops" to be grown simultaneously. By staggering sowings, a constant supply of young sprouts can be ensured. Any vessel used for sprouting must allow water to drain from it, because sprouts that sit in water will rot quickly. The seeds swell, may stick to the sides of the jar, and begin germinating within a day or two.

Another sprouting technique is to use a pulse drip method. The photo shows crimson clover sprouts grown on 1/8" urethane foam mats. It's a one-way watering system with micro sprinklers providing intermittent pulses of fresh water to reduce the risk of bacterial cross-contamination with Salmonella and E. coli during the sprouting process.

Crimson clover sprouts grown on 1/8" urethane foam mats with a pulse drip technique. Four micro-sprinklers cycle pulsing continuously over a 7-day period, each putting out about 1/2 gallon per hour. The four micro-sprinklers were each fitted with an LPD to keep the lines fully charged between pulses.

Sprouts are rinsed two to four times a day, depending on the climate and the type of seed, to provide them with moisture and prevent them from souring. Each seed has its own ideal sprouting time. After three to five days the sprouts will have grown to 5 to 8 centimetres (2–3 in) in length and will be suitable for consumption. If left longer they will begin to develop leaves, and are then known as baby greens. A popular baby green is sunflower after 7–10 days. Refrigeration can be used as needed to slow or halt the growth process of any sprout.

Common causes for sprouts to become inedible:

- Seeds are not rinsed well enough before soaking
- Seeds are left in standing water after the initial soaking
- Seeds are allowed to dry out
- Temperature is too high or too low
- Dirty equipment
- Insufficient air flow
- Contaminated water source
- Poor germination rate

Mung beans can be sprouted either in light or dark conditions. Those sprouted in the dark will be crisper in texture and whiter, as in the case of commercially available Chinese Bean Sprouts, but these have less nutritional content than those grown in partial sunlight. Growing in full sunlight is not recommended, because it can cause the beans to overheat or dry out. Subjecting the sprouts to pressure, for example, by placing a weight on top of them in their sprouting container, will result in larger, crunchier sprouts similar to those sold in grocery stores.

A very effective way to sprout beans like lentils or azuki is in colanders. Soak the beans in water for about 8 hours then place in the colander. Wash twice a day. The sprouted beans can be eaten raw or cooked.

Sprouting is also applied on a large scale to barley as a part of the malting process. Malted barley is an important ingredient in beer and is used in huge quantities. Most malted barley is distributed among wide retail sellers in North American regions.

Many varieties of nuts, such as almonds and peanuts, can also be started in their growth cycle by soaking and sprouting, although because the sprouts are generally still very small when eaten, they are usually called “soaks”.

Nutritional Information

Sprouts used for a verrine.

Fresh mung bean sprouts in a bowl

Sprouts are said to be rich in digestible energy, bioavailable vitamins, minerals, amino acids, proteins, and phytochemicals, as these are necessary for a germinating plant to grow. These nutrients are essential for human health. The nutritional changes upon germination and sprouting are summarised below.

Chavan and Kadam (1989) concluded that

- "The desirable nutritional changes that occur during sprouting are mainly due to the breakdown of complex compounds into a more simple form, transformation into essential constituents and breakdown of nutritionally undesirable constituents."
- "The metabolic activity of resting seeds increases as soon as they are hydrated during soaking. Complex biochemical changes occur during hydration and subsequent sprouting. The reserve chemical constituents, such as protein, starch and lipids, are broken down by enzymes into simple compounds that are used to make new compounds."
- "Sprouting grains causes increased activities of hydrolytic enzymes, improvements in the contents of total proteins, fat, certain essential amino acids, total sugars, B-group vitamins, and a decrease in dry matter, starch and anti-nutrients. The increased contents of protein, fat, fibre and total ash are only apparent and attributable to the disappearance of starch. However, improvements in amino acid composition, B-group vitamins, sugars, protein and starch digestibilities, and decrease in phytates and protease inhibitors are the metabolic effects of the sprouting process."

Increases in Protein Quality Chavan and Kadam (1989) stated that "Very complex qualitative changes are reported to occur during soaking and sprouting of seeds. The conversion of storage proteins of cereal grains into albumins and globulins during sprouting may improve the quality of cereal proteins. Many studies have shown an increase in the content of the amino acid Lysine with sprouting."

"An increase in proteolytic activity during sprouting is desirable for nutritional improvement of cereals because it leads to hydrolysis of prolamins and the liberated amino acids such as glutamic and proline are converted to limiting amino acids such as lysine."

Increases in Crude Fibre content Cuddeford (1989), based on data obtained by Peer and Leeson (1985) stated that "In sprouted barley, crude fibre, a major constituent of cell walls, increases both in percentage and real terms, with the synthesis of structural carbohydrates, such as cellulose and hemicellulose". Chung et al. (1989) found that the fibre content increased from 3.75% in unsprouted barley seed to 6% in 5-day sprouts."

Crude Protein and Crude Fibre changes in Barley Sprouted over a 7-day period

	Crude Protein (% of DM)	Crude Fibre (% of DM)
Original seed	12.7%	5.4%
Day 1	12.7%	5.6%
Day 2	13.0%	5.9%
Day 3	13.6%	5.8%
Day 4	13.4%	7.4%
Day 5	13.9%	9.7%
Day 6	14.0%	10.8%
Day 7	15.5%	14.1%

Source: Cuddeford (1989), based on data obtained by Peer and Leeson (1985).

Increase of protein is not due to new protein being manufactured by the germination process but by the washing out of starch and conversion to fiber—increasing the relative proportion of protein.

Increases in Essential Fatty Acids

An increase in lipase activity has been reported in barley by MacLeod and White (1962), as cited by Chavan and Kadam (1989). Increased lipolytic activity during germination and sprouting causes hydrolysis of triacylglycerols to glycerol and constituent fatty acids.

Increases in Vitamin content According to Chavan and Kadam (1989), most reports agree that sprouting treatment of cereal grains generally improves their vitamin value, especially the B-group vitamins. Certain vitamins such as α-tocopherol (Vitamin-E) and β-carotene (Vitamin-A precursor) are produced during the growth process (Cuddeford, 1989).

According to Shipard (2005), "Sprouts provide a good supply of Vitamins A, E & C plus B complex. Like enzymes, vitamins serve as bioactive catalysts to assist in the digestion and metabolism of feeds and the release of energy. They are also essential for the healing and repair of cells. However, vitamins are very perishable, and in general, the fresher the feeds eaten, the higher the vitamin content. The vitamin content of some seeds can increase by up to 20 times their original value within several days of sprouting. Mung Bean sprouts have B vitamin increases, compared to the dry seeds, of - B_1 up 285%, B_2 up 515%, B_3 up 256%. Even soaking seeds overnight in water yields greatly increased amounts of B vitamins, as well as Vitamin C. Compared with mature plants, sprouts can yield vitamin contents 30 times higher."

Chelation of Minerals Shipard (2005) claims that - "When seeds are sprouted, minerals chelate or merge with protein, in a way that increases their function."

It is important to note that while these changes may sound impressive, the comparisons are between dormant non-sprouted seed to sprouted seed rather than comparisons of sprouts to mature vegetables. Compared to dry seeds there are very large increases in nutrients whereas compared with mature vegetables the increase is less. However, a sprout, just starting out in life, is likely to need and thus have more nutrients (percentage wise) than a mature vegetable.

Following table lists selected nutrients in kidney beans to show the effect of sprouting. Nutrients are calculated for 100 gms of non-water components to take the water out of equation since after sprouting beans absorb lot of water.

GABA Sprouting also have shown that can improve the levels of GABA, a compound involved in the regulation of blood pressure, and promoted the liberation of bioactive peptides in diverse legumes.

Nutrients in grams per 100 grams of non-water components of kidney beans				
Nutrient	**DRI**	**Raw Beans**	**Sprouted Beans**	**Ratio**
Protein (g)	50	26.72	45.15	1.7
Vitamin C (mg)	90	5.1	416	81.56
Thiamin (mg)	1.2	0.6	3.98	6.63
Riboflavin (mg)	1.3	.248	2.78	11.2

Nutrients in grams per 100 grams of non-water components of kidney beans				
Nutrient	**DRI**	**Raw Beans**	**Sprouted Beans**	**Ratio**
Niacin (mg)	16	2.33	31.40	13.47
Vitamin B-6 (mg)	1.3	0.45	0.91	2.02
Folate (μg)	400	446	634.25	1.422
Vitamin E(mg)	15	.25	0	0
Vitamin K (μg)	120	21.527	0	0

Health Concerns

Bacterial Infection

FDA health warning on a sprouts package

Commercially grown sprouts have been associated with multiple outbreaks of harmful bacteria, including salmonella and toxic forms of *Escherichia coli*. Such infections may be a result of contaminated seeds or of unhygienic production with high microbial counts. Sprout seeds can become contaminated in the fields where they are grown, and sanitizing steps may be unable to kill bacteria hidden in damaged seeds. A single surviving bacterium in a kilogram of seed can be enough to contaminate a whole batch of sprouts, according to the FDA.

To minimize the impact of the incidents and maintain public health, both the U.S. Food and Drug Administration (FDA) and Health Canada issued industry guidance on the safe manufacturing of edible sprouts and public education on their safe consumption. There are also publications for hobby farmers on safely growing and consuming sprouts at home. The recommendations include development and implementation of good agricultural practices and good manufacturing practices in the production and handling of seeds and sprouts, seed disinfection treatments, and microbial testing before the product enters the food supply.

In June 2011, contaminated fenugreek sprouts (grown from seed from Egypt) in Germany was identified as the source of the 2011 E. coli O104:H4 outbreak which the German officials had blamed wrongly, first on cucumbers from Spain and then on mung bean sprouts. In addition to Germany, where 3,785 cases and 45 deaths had been reported by the end of the outbreak, a handful of cases were reported in several countries including Switzerland, Poland, the Netherlands, Sweden, Denmark, the UK, Canada and the USA. Virtually all affected people had been in Germany shortly before becoming ill.

Antinutritional Factors

Some legumes, including sprouts, can contain toxins or antinutritional factors, which can be reduced by soaking, sprouting and cooking (e.g., stir frying). Joy Larkcom advises that to be on the safe side "one shouldn't eat large quantities of raw legume sprouts on a regular basis, no more than about 550g (20oz) daily".

Phytic acid, an antinutritional factor, occurs primarily in the seed coats and germ tissue of plant seeds. It forms insoluble or nearly insoluble compounds with many metal ions, including those of calcium, iron, magnesium and zinc, reducing their dietary availability. Diets high in phytic acid content and poor in these minerals produce mineral deficiency in experimental animals (Gontzea and Sutzescu, 1968, as cited in Chavan and Kadam, 1989). The latter authors state that the sprouting of cereals has been reported to decrease levels of phytic acid. Similarly, Shipard (2005) states that enzymes of germination and sprouting can help eliminate detrimental substances such as phytic acid. However, the amount of phytic acid reduction from soaking is only marginal, and not enough to counteract its antinutrient effects.

Standards and Regulations

EU Regulation

In order to prevent incidents like the 2011 EHEC epidemic, the European Commission has issued three new, tightened regulations on March 11, 2013.

- Regulation (EU) No 208/2013*

The origin of the seeds has to be traceable always at all stages of processing, production and distribution. Therefore, a full description of the seeds or sprouts needs to be kept on record.(Article 18 of Regulation (EC) No 178/2002)

- Regulation (EU) No 209/2013*

This regulation amends Regulation (EC) No 2073/2005 in respect of microbiological criteria for sprouts and the sampling rules for poultry carcases and fresh poultry meat.

- Regulation (EU) No 211/2013*

Imported sprouts or seeds intended for the production of sprouts need a certificate according to the model declared in the Annex of this regulation. The certificate serves as proof that the production process complies with the general hygiene provisions in Part A of Annex I to Regulation (EC) No 852/2004 and the traceability requirements of Implementing Regulation (EU) No 208/2013.

References

- Cain M.D., Shelton M.G. (2001). "Twenty years of natural loblolly and shortleaf pine seed production on the Crossett Experimental Forest in southeastern Arkansas". Southern Journal of Applied Forestry. 25 (1): 40–45
- Rost, Thomas L.; Weier, T. Elliot; Weier, Thomas Elliot (1979). Botany: a brief introduction to plant biology. New York: Wiley. p. 319. ISBN 0-471-02114-8

- Jon E. Keeley and Fotheringham. "Trace Gas Emissions and Smoke-Induced Seed Germination". Science. 276 (276 (5316)): 1248. doi:10.1126/science.276.5316.1248
- Vleeshouwers L.M., Bouwmeester H.J., Karssen C.M. (1995). "Redefining seed dormancy: an attempt to integrate physiology and ecology". Journal of Ecology. 83: 1031–1037. doi:10.2307/2261184
- "Science Magazine". Runcaria, a Middle Devonian Seed Plant Precursor. American Association for the Advancement of Science. 2011. Retrieved March 22, 2011
- Sabelli, P.A.; Larkins, B.A. (2009). "The Development of Endosperm in Grasses". Plant Physiology. 149 (1): 14–26. doi:10.1104/pp.108.129437. PMC 2613697. PMID 19126691
- Patten D.T. (1978). "Productivity and production efficiency of an Upper Sonoran Desert ephemeral community". American Journal of Botany. 65 (8): 891–895. doi:10.2307/2442185
- Morhardt, Sia; Morhardt, Emil; Emil Morhardt, J. (2004). California desert flowers: an introduction to families, genera, and species. Berkeley: University of California Press. p. 24. ISBN 0-520-24003-0
- Cody, M.L. & Overton, J.M. (1996). "Short-term evolution of reduced dispersal in island plant populations". Journal of Ecology. 84: 53–61. doi:10.2307/2261699. JSTOR 2261699
- Wedin GP; Neal JS; Everson GW; Krenzelok EP (May 1986). "Castor bean poisoning". Am J Emerg Med. 4 (3): 259–61. doi:10.1016/0735-6757(86)90080-X. PMID 3964368
- Hartmann, Hudson Thomas, and Dale E. Kester. 1983. Plant propagation principles and practices. Englewood Cliffs, N.J.: Prentice-Hall. ISBN 0-13-681007-1. Pages 175-77
- Fedriani, J. M., Delibes, M. (2009). "Functional diversity in fruit-frugivore interactions: A field experiment with Mediterranean mammals". Ecography. 32 (6): 983. doi:10.1111/j.1600-0587.2009.05925.x
- Palmer, Jeffrey D.; Soltis, Douglas E.; Chase, Mark W. (2004). "The plant tree of life: an overview and some points of view". American Journal of Botany. 91 (10): 1437–1445. doi:10.3732/ajb.91.10.1437. PMID 21652302
- Taylor EL; Taylor TMC (1993). The biology and evolution of fossil plants. Englewood Cliffs, N.J: Prentice Hall. p. 466. ISBN 0-13-651589-4
- Ran, Nathan (August 11, 2006). "Long-Distance Dispersal of Plants". Science. 313 (5788): 786–788. Bibcode:2006Sci...313..786N. doi:10.1126/science.1124975
- James A. Doyle (January 2006). "Seed ferns and the origin of angiosperms". The Journal of the Torrey Botanical Society. 133 (1): 169–209. doi:10.3159/1095-5674(2006)133[169:SFATOO]2.0.CO;2. ISSN 1095-5674
- Schulze, Ernst-Detlef; Beck, Erwin & Müller-Hohenstein, Klaus (2005). Plant Ecology. Springer. pp. 543–. ISBN 978-3-540-20833-4
- Igor Kosinki (2007). "Long-term variability in seed size and seedling establishment of Maianthemum bifolium". Plant Ecology. 194 (2): 149–156. doi:10.1007/s11258-007-9281-1
- Soltis, Douglas E.; Soltis, Pamela S.; Zanis, Michael J. (2002). "Phylogeny of seed plants based on evidence from eight genes". American Journal of Botany. 89 (10): 1670–1681. doi:10.3732/ajb.89.10.1670. PMID 21665594
- Novíkov & Barabaš-Krasni (2015). "Modern plant systematics". Liga-Pres: 685. doi:10.13140/RG.2.1.4745.6164. ISBN 978-966-397-276-3
- Franceschini, M. (2004) "An unusual case of epigeal cryptocotylar germination in Rollinia salicifolia (Annonaceae)" Botanical Journal of the Linnean Society vol. 146 no. 1
- Anderson, Anderson & Cleal (2007). "Brief history of the gymnosperms: classification, biodiversity, phytogeography and ecology". Strelitzia. SANBI. 20: 280. ISBN 978-1-919976-39-6
- Thompson K, Ceriani RM, Bakker JP, Bekker RM (2003). "Are seed dormancy and persistence in soil related?". 13: 97–100. doi:10.1079/ssr2003128

3

Pollination: An Overview

The transfer of pollen from the male to the female part of the plant is known as pollination. There are two types of pollinations that occur, self-pollination and cross-pollination (xenogamy). A pollinator is an animal that aids in a plant's pollination. This chapter is an overview of the subject matter incorporating all the main aspects of pollination.

Pollination

Pollination is the process by which pollen is transferred to the female reproductive organs of a plant, thereby enabling fertilization to take place. Like all living organisms, seed plants have a single major goal: to pass their genetic information on to the next generation. The reproductive unit is the seed, and pollination is an essential step in the production of seeds in all spermatophytes (seed plants).

Carpenter bee with pollen collected from Night-blooming cereus

For the process of pollination to be successful, a pollen grain produced by the anther, the male part of a flower, must be transferred to a stigma, the female part of the flower, of a plant of the same species. The process is rather different in angiosperms (flowering plants) from what it is in gymnosperms (other seed plants). In angiosperms, after the pollen grain has landed on the stigma, it creates a pollen tube which grows down the style until it reaches the ovary. Sperm cells from the pollen grain then move along the pollen tube, enter the egg cell through the micropyle and fertilise it, resulting in the production of a seed.

Tip of a tulip stamen covered with pollen grains.

A successful angiosperm pollen grain (gametophyte) containing the male gametes is transported to the stigma, where it germinates and its pollen tube grows down the style to the ovary. Its two gametes travel down the tube to where the gametophyte(s) containing the female gametes are held within the carpel. One nucleus fuses with the polar bodies to produce the endosperm tissues, and the other with the ovule to produce the embryo Hence the term: "double fertilization".

In gymnosperms, the ovule is not contained in a carpel, but exposed on the surface of a dedicated support organ, such as the scale of a cone, so that the penetration of carpel tissue is unnecessary. Details of the process vary according to the division of gymnosperms in question. Two main modes of fertilization are found in gymnosperms. Cycads and *Ginkgo* have motile sperm that swim directly to the egg inside the ovule, whereas conifers and gnetophytes have sperm that are unable to swim but are conveyed to the egg along a pollen tube.

The study of pollination brings together many disciplines, such as botany, horticulture, entomology, and ecology. The pollination process as an interaction between flower and pollen vector was first addressed in the 18th century by Christian Konrad Sprengel. It is important in horticulture and agriculture, because fruiting is dependent on fertilization: the result of pollination. The study of pollination by insects is known as *anthecology*.

Pollination Process

Pollen germination has three stages; hydration, activation and pollen tube emergence. The pollen grain is severely dehydrated so that its mass is reduced enabling it to be more easily transported from flower to flower. Germination only takes place after rehydration, ensuring that premature germination does not take place in the anther. Hydration allows the plasma membrane of the pollen grain to reform into its normal bilayer organization providing an effective osmotic membrane. Activation involves the development of actin filaments throughout the cytoplasm of the cell, which eventually become concentrated at the point from which the pollen tube will emerge. Hydration and activation continue as the pollen tube begins to grow.

In conifers, the reproductive structures are borne on cones. The cones are either pollen cones (male) or ovulate cones (female), but some species are monoecious and others dioecious. A pollen cone contains hundreds of microsporangia carried on (or borne on) reproductive structures called sporophylls. Spore mother cells in the microsporangia divide by meiosis to form haploid microspores that develop further by two mitotic divisions into immature male gametophytes (pollen grains). The four resulting cells consist of a large tube cell that forms the pollen tube, a generative cell that will produce two sperm by mitosis, and two prothallial cells that degenerate. These cells comprise a very reduced microgametophyte, that is contained within the resistant wall of the pollen grain.

The pollen grains are dispersed by the wind to the female, ovulate cone that is made up of many overlapping scales (sporophylls, and thus megasporophylls), each protecting two ovules, each of which consists of a megasporangium (the nucellus) wrapped in two layers of tissue, the integument and the cupule, that were derived from highly modified branches of ancestral gymnosperms. When a pollen grain lands close enough to the tip of an ovule, it is drawn in through the micropyle (a pore in the integuments covering the tip of the ovule) often by means of a drop of liquid known as a pollination drop. The pollen enters a pollen chamber close to the nucellus, and there it may wait for a year before it germinates and forms a pollen tube that grows through the wall of the megasporangium (=nucellus) where fertilisation takes place. During this time, the megaspore mother cell divides by meiosis to form four haploid cells, three of which degenerate. The surviving one develops as a megaspore and divides repeatedly to form an immature female gametophyte (egg sac). Two or three archegonia containing an egg then develop inside the gametophyte. Meanwhile, in the spring of the second year two sperm cells are produced by mitosis of the body cell of the male gametophyte. The pollen tube elongates and pierces and grows through the megasporangium wall and delivers the sperm cells to the female gametophyte inside. Fertilisation takes place when the nucleus of one of the sperm cells enters the egg cell in the megagametophyte's archegonium.

In flowering plants, the anthers of the flower produce microspores by meiosis. These undergo mitosis to form male gametophytes, each of which contains two haploid cells. Meanwhile, the ovules produce megaspores by meiosis, further division of these form the female gametophytes, which are very strongly reduced, each consisting only of a few cells, one of which is the egg. When a pollen grain adheres to the stigma of a carpel it germinates, developing a pollen tube that grows through the tissues of the style, entering the ovule through the micropyle. When the tube reaches the egg sac, two sperm cells pass through it into the female gametophyte and fertilisation takes place.

Types

Depending on the source of pollen, pollination can be classified into 2 types - Self-pollination and Cross Pollination (Xenogamy). Self Pollination is further divided into Autogamy and Geitonogamy. Depending on agent of Pollination, pollination can be classified into abiotic pollination and biotic pollination.

Self Pollination is the type of Pollination in which pollen grains are transferred from anther to the stigma of the same flower (Autogamy) or pollen grains are transferred from anther to the stigma of different flower of the same plant (Geitonogamy).

Cross Pollination or Xenogamy is the type of pollination in which pollen grains are transferred from anther to the stigma of a different plant.

Abiotic

Abiotic pollination refers to situations where pollination is mediated without the involvement of other organisms. The most common form of abiotic pollination, anemophily, is pollination by wind. Wind pollination is very imprecise, with a minute proportion of pollen grains landing by chance on a suitable receptive stigma, the rest being wasted in the environment. This form of pollination is used by grasses, most conifers, and many deciduous trees. Hydrophily is pollination by water, and occurs in aquatic plants which release their pollen directly into the surrounding water. About 80% of all plant pollination is biotic. In gymnosperms, biotic pollination is generally incidental when it occurs, though some gymnosperms and their pollinators are mutually adapted for pollination. The best-known examples probably are members of the order Cycadales and associated species of beetles. Of the abiotically pollinated species of plant, 98% are anemophilous and 2% hydrophilous, their pollen being transported by water.

It is thought that among angiosperms, entomophily is the primitive state; this is indicated by the vestigial nectaries in the wind-pollinated *Urtica* and other plants, and the presence of fragrances in some of these plants. Of the angiosperms, grasses, sedges, rushes and catkin-bearing plants are in general wind pollinated. Other flowering plants are mostly biotic, the pollen being carried by animal vectors. However a number of plants in multiple families have secondarily adopted wind pollination in contrast to other members of their groups. Some plants are intermediate between the two pollination methods. common heather is regularly pollinated by insects, but produce clouds of pollen and some wind pollination is inevitable, and the hoary plantain is primarily wind pollinated, but is also visited by insects which pollinate it.

Melissodes desponsa covered in pollen

Biotic

Hummingbirds typically feed on red flowers

More commonly, the process of pollination requires pollinators: organisms that carry or move the pollen grains from the anther of one flower to the receptive part of the carpel or pistil (stigma) of another. This is biotic pollination. The various flower traits (and combinations thereof) that differentially attract one type of pollinator or another are known as pollination syndromes. At least 100,000 species of animal, and possibly as many as 200,000, act as pollinators of the estimated 250,000 species of flowering plants in the world. The majority of these pollinators are insects, but about 1,500 species of birds and mammals have been reported to visit flowers and may transfer pollen between them. Besides birds and bats which are the most frequent visitors, these include monkeys, lemurs, squirrels, rodents and possums.

Entomophily, pollination by insects, often occurs on plants that have developed colored petals and a strong scent to attract insects such as, bees, wasps and occasionally ants (Hymenoptera), beetles (Coleoptera), moths and butterflies (Lepidoptera), and flies (Diptera). The existence of insect pollination dates back to the dinosaur era.

In zoophily, pollination is performed by vertebrates such as birds and bats, particularly, hummingbirds, sunbirds, spiderhunters, honeyeaters, and fruit bats. Ornithophily or bird pollination is the pollination of flowering plants by birds. Chiropterophily or bat pollination is the pollination of flowering plants by bats. Plants adapted to use bats or moths as pollinators typically have white petals, strong scent and flower at night, whereas plants that use birds as pollinators tend to produce copious nectar and have red petals.

Insect pollinators such as honeybees (*Apis mellifera*), bumblebees (*Bombus terrestris*), and butterflies (*Thymelicus flavus*) have been observed to engage in flower constancy, which means they are more likely to transfer pollen to other conspecific plants. This can be beneficial for the pollinators, as flower constancy prevents the loss of pollen during interspecific flights and pollinators from clogging stigmas with pollen of other flower species. It also improves the probability that the pollinator will find productive flowers easily accessible and recognisable by familiar clues.

Mechanism

A European honey bee collects nectar, while pollen collects on its body.

Africanized honey bees immersed in Yellow Opuntia engelmannii Cactus Flower Pollen

Diadasia Bee Straddles Cactus Flower Carpels

Pollination can be accomplished by cross-pollination or by self-pollination:

- Cross-pollination, also called *allogamy*, occurs when pollen is delivered from the stamen of one flower to the stigma of a flower on another plant of the same species. Plants adapted for cross-pollination have several mechanisms to prevent self-pollination; the reproductive organs may be arranged in such a way that self-fertilisation is unlikely, or the stamens and carpels may mature at different times.

- Self-pollination occurs when pollen from one flower pollinates the same flower or other flowers of the same individual. It is thought to have evolved under conditions when pollinators were not reliable vectors for pollen transport, and is most often seen in short-lived annual species and plants that colonize new locations. Self-pollination may include *autogamy*, where pollen is transferred to the female part of the same flower; or *geitonogamy*, when pollen is transferred to another flower on the same plant. Plants adapted to self-fertilize often have similar stamen and carpel lengths. Plants that can pollinate themselves and produce viable offspring are called self-fertile. Plants that cannot fertilize themselves are called self-sterile, a condition which mandates cross-pollination for the production of offspring.

- *Cleistogamy*: is self-pollination that occurs before the flower opens. The pollen is released from the anther within the flower or the pollen on the anther grows a tube down the style to the ovules. It is a type of sexual breeding, in contrast to asexual systems such as apomixis.

Some *cleistogamous* flowers never open, in contrast to *chasmogamous* flowers that open and are then pollinated. Cleistogamous flowers are by necessity found on self-compatible or self-fertile plants. Although certain orchids and grasses are entirely cleistogamous, other plants resort to this strategy under adverse conditions. Often there may be a mixture of both cleistogamous and chasmogamous flowers, sometimes on different parts of the plant and sometimes in mixed inflorescences. The ground bean produces cleistogamous flowers below ground, and mixed cleistogamous and chasmogamous flowers above.

Geranium incanum, like most geraniums and pelargoniums, sheds its anthers, sometimes its stamens as well, as a barrier to self-pollination. This young flower is about to open its anthers, but has not yet fully developed its pistil.

These *Geranium incanum* flowers have opened their anthers, but not yet their stigmas. Note the change of colour that signals to pollinators that it is ready for visits.

This *Geranium incanum* flower has shed its stamens, and deployed the tips of its pistil without accepting pollen from its own anthers. (It might of course still receive pollen from younger flowers on the same plant.)

An estimated 48.7% of plant species are either dioecious or self-incompatible obligate out-crossers. It is also estimated that about 42% of flowering plants exhibit a mixed mating system in nature. In the most common kind of mixed mating system, individual plants produce a single type of flower and fruits may contain self-pollinated, out-crossed or a mixture of progeny types.

Pollination also requires consideration of pollenizers. The terms "pollinator" and "pollenizer" are often confused: a *pollinator* is the agent that moves the pollen, whether it be bees, flies, bats, moths, or birds; a *pollenizer* is the plant that serves as the pollen source for other plants. Some plants are *self-compatible* (*self-fertile*) and can pollinate and fertilize themselves. Other plants have chemical or physical barriers to self-pollination.

In agriculture and horticulture pollination management, a good pollenizer is a plant that provides compatible, viable and plentiful pollen and blooms at the same time as the plant that is to be pollinated or has pollen that can be stored and used when needed to pollinate the desired flowers. Hybridization is effective pollination between flowers of different species, or between different breeding lines or populations.

Peaches are considered self-fertile because a commercial crop can be produced without cross-pollination, though cross-pollination usually gives a better crop. Apples are considered self-incompatible, because a commercial crop must be cross-pollinated. Many commercial fruit tree varieties are grafted clones, genetically identical. An orchard block of apples of one variety is genetically a single plant. Many growers now consider this a mistake. One means of correcting this mistake is to graft a limb of an appropriate pollenizer (generally a variety of crabapple) every six trees or so.

The wasp *Mischocyttarus* rotundicollis transporting pollen grains of *Schinus terebinthifolius*

Pollen Vectors

Biotic pollen vectors are animals, usually insects, but also reptiles, birds, mammals, and sundry others, that routinely transport pollen and play a role in pollination. This is usually as a result of their activities when visiting plants for feeding, breeding or shelter. The pollen adheres to the vector's body parts such as face, legs, mouthparts, hair, feathers, and moist spots; depending on the particular vector. Such transport is vital to the pollination of many plant species.

Any kind of animal that often visits or encounters flowers is likely to be a pollen vector to some extent. For example, a crab spider that stops at one flower for a time and then moves on, might carry pollen incidentally, but most pollen vectors of significant interest are those that routinely visit the flowers for some functional activity. They might feed on pollen, or plant organs, or on plant secretions such as nectar, and carry out acts of pollination on the way. Many plants bear flowers that favour certain types of pollinator over all others. This need not always be an effective strategy, because some flowers that are of such a shape that they favor pollinators that pass by their anthers and stigmata on the way to the nectar, may get robbed by ants that are small enough to bypass the normal channels, or by short-tongued bees that bite through the bases of deep corolla tubes to extract nectar at the end opposite to the anthers and stigma. Some flowers have specialized mechanisms to trap pollinators to increase effectiveness. Other flowers will attract pollinators by odor. For example, bee species such as *Euglossa cordata* are attracted to orchids this way, and it has been suggested that the bees will become intoxicated during these visits to the orchid flowers, which last up to 90 minutes. However, in general, plants that rely on pollen vectors tend to be adapted to their particular type of vector, for example day-pollinated species tend to be brightly coloured, but if they are pollinated largely by birds or specialist mammals, they tend to be larger and have larger nectar rewards than species that are strictly insect-pollinated. They also tend to spread their rewards over longer periods, having long flowering seasons; their specialist pollinators would be likely to starve if the pollination season were too short.

As for the types of pollinators, reptile pollinators are known, but they form a minority in most ecological situations. They are most frequent and most ecologically significant in island systems, where insect and sometimes also bird populations may be unstable and less species-rich. Adaptation to a lack of animal food and of predation pressure, might therefore favour reptiles becoming more herbivorous and more inclined to feed on pollen and nectar. Most species of lizards in the families that seem to be significant in pollination seem to carry pollen only incidentally, especially the larger species such as Varanidae and Iguanidae, but especially several species of the Gekkonidae are active pollinators, and so is at least one species of the Lacertidae, *Podarcis lilfordi*, which pollinates various species, but in particular is the major pollinator of *Euphorbia dendroides* on various Mediterranean islands.

Mammals are not generally thought of as pollinators, but some rodents, bats and marsupials are significant pollinators and some even specialise in such activities. In South Africa certain species of *Protea* (in particular *Protea humiflora*, *P. amplexicaulis*, *P. subulifolia*, *P. decurrens* and *P. cordata*) are adapted to pollination by rodents (particularly Cape Spiny Mouse, *Acomys subspinosus*) and elephant shrews (*Elephantulus* species). The flowers are borne near the ground, are yeasty smelling, not colourful, and sunbirds reject the nectar with its high xylose content. The mice apparently can digest the xylose and they eat large quantities of the pollen. In Australia pollination by flying, gliding and earthbound mammals has been demonstrated.

Examples of pollen vectors include many species of wasps, that transport pollen of many plant species, being potential or even efficient pollinators.

Evolution of Plant/pollinator Interactions

The first fossil record for abiotic pollination is from fern-like plants in the late Carboniferous period. Gymnosperms show evidence for biotic pollination as early as the Triassic period. Many fossilized pollen grains show characteristics similar to the biotically dispersed pollen today. Furthermore, the gut contents, wing structures, and mouthpart morphologies of fossilized beetles and flies suggest that they acted as early pollinators. The association between beetles and angiosperms during the early Cretaceous period led to parallel radiations of angiosperms and insects into the late Cretaceous. The evolution of nectaries in late Cretaceous flowers signals the beginning of the mutualism between hymenopterans and angiosperms.

Bees provide a good example of the mutualism that exists between hymenopterans and angiosperms. Flowers provide bees with nectar (an energy source) and pollen (a source of protein). When bees go from flower to flower collecting pollen they are also depositing pollen grains onto the flowers, thus pollinating them. While pollen and nectar, in most cases, are the most notable reward attained from flowers, bees also visit flowers for other resources such as oil, fragrance, resin and even waxes. It has been estimated that bees originated with the origin or diversification of angiosperms. In addition, cases of coevolution between bee species and flowering plants have been illustrated by specialized adaptations. For example, long legs are selected for in *Rediviva neliana*, a bee that collects oil from *Diascia capsularis*, which have long spur lengths that are selected for in order to deposit pollen on the oil-collecting bee, which in turn selects for even longer legs in *R. neliana* and again longer spur length in *D. capsularis* is selected for, thus, continually driving each other's evolution.

In Agriculture

An *Andrena* bee collects pollen among the stamens of a rose. The female carpel structure appears rough and globular to the left. The bee's stash of pollen is on its hind leg.

Bombus ignitus, a popular commercial pollinator in Japan and China

Blueberries being pollinated by bumblebees. Bumblebee hives need to be bought each year as the queens must hibernate (unlike honey bees). They are used nonetheless as they offer advantages with certain fruits as blueberries (such as the fact that they are active even at colder outdoor ambient temperature).

Well-pollinated blackberry blossom begins to develop fruit. Each incipient drupelet has its own stigma and good pollination requires the delivery of many grains of pollen to the flower so that all drupelets develop.

Pollination management is a branch of agriculture that seeks to protect and enhance present pollinators and often involves the culture and addition of pollinators in monoculture situations, such as commercial fruit orchards. The largest managed pollination event in the world is in Californian almond orchards, where nearly half (about one million hives) of the US honey bees are trucked to the almond orchards each spring. New York's apple crop requires about 30,000 hives; Maine's blueberry crop uses about 50,000 hives each year.

Bees are also brought to commercial plantings of cucumbers, squash, melons, strawberries, and many other crops. Honey bees are not the only managed pollinators: a few other species of bees are also raised as pollinators. The alfalfa leafcutter bee is an important pollinator for alfalfa seed in western United States and Canada. Bumblebees are increasingly raised and used extensively for greenhouse tomatoes and other crops.

The ecological and financial importance of natural pollination by insects to agricultural crops, improving their quality and quantity, becomes more and more appreciated and has given rise to new financial opportunities. The vicinity of a forest or wild grasslands with native pollinators near agricultural crops, such as apples, almonds or coffee can improve their yield by about 20%. The benefits of native pollinators may result in forest owners demanding payment for their contribution in the improved crop results – a simple example of the economic value of ecological services. Farmers can also raise native crops in order to promote native bee pollinator species as shown with *L. vierecki* in Delaware and *L. leucozonium* in southwest Virginia.

The American Institute of Biological Sciences reports that native insect pollination saves the United States agricultural economy nearly an estimated $3.1 billion annually through natural crop production; pollination produces some $40 billion worth of products annually in the United States alone.

Pollination of food crops has become an environmental issue, due to two trends. The trend to monoculture means that greater concentrations of pollinators are needed at bloom time than ever before, yet the area is forage poor or even deadly to bees for the rest of the season. The other trend is the decline of pollinator populations, due to pesticide misuse and overuse, new diseases and parasites of bees, clearcut logging, decline of beekeeping, suburban development, removal of hedges and other habitat from farms, and public concern about bees. Widespread aerial spraying for mosquitoes due to West Nile fears is causing an acceleration of the loss of pollinators.

The US solution to the pollinator shortage, so far, has been for commercial beekeepers to become pollination contractors and to migrate. Just as the combine harvesters follow the wheat harvest from Texas to Manitoba, beekeepers follow the bloom from south to north, to provide pollination for many different crops.

In some situations, farmers or horticulturists may aim to restrict natural pollination to only permit breeding with the preferred individuals plants. This may be achieved through the use of pollination bags.

Improving Pollination in Areas with Suboptimal Bee Densities

In some instances growers' demand for beehives far exceeds the available supply. The number of managed beehives in the US has steadily declined from close to 6 million after WWII, to less than 2.5 million today. In contrast, the area dedicated to growing bee-pollinated crops has grown over 300% in the same time period. Additionally, in the past five years there has been a decline in winter managed beehives, which has reached an unprecedented rate of colony losses at near 30%. At present, there is an enormous demand for beehive rentals that cannot always be met. There is a clear need across the agricultural industry for a management tool to draw pollinators into cultivations and encourage them to preferentially visit and pollinate the flowering crop. By attracting pollinators like honeybees and increasing their foraging behavior, particularly in the center of large plots, we can increase grower returns and optimize yield from their plantings. ISCA Technologies, from Riverside California, created a semiochemical formulation called SPLAT Bloom, that modifies the behavior of honeybees, inciting them to visit flowers in every portion of the field.

Environmental Impacts

Loss of pollinators, also known as Pollinator decline (of which colony collapse disorder is perhaps the most well known) has been noticed in recent years. Observed losses would have significant economic impacts. Possible explanations for pollinator decline include habitat destruction, pesticide, parasitism/diseases, climate change and others, and many researchers believe it is the synergistic effects of these factors which are ultimately detrimental to pollinator populations.

The Structure of Plant–pollinator Networks

Wild pollinators often visit a large number of plant species and plants are visited by a large number of pollinator species. All these relations together form a network of interactions between plants and pollinators. Surprising similarities were found in the structure of networks consisting out of the interactions between plants and pollinators. This structure was found to be similar in very different ecosystems on different continents, consisting of entirely different species.

The structure of plant-pollinator networks may have large consequences for the way in which pollinator communities respond to increasingly harsh conditions. Mathematical models, examining the consequences of this network structure for the stability of pollinator communities suggest that the specific way in which plant-pollinator networks are organized minimizes competition between pollinators and may even lead to strong indirect facilitation between pollinators when conditions are harsh. This means that pollinator species together can survive under harsh conditions. But it also means that pollinator species collapse simultaneously when conditions pass a critical point. This simultaneous collapse occurs, because pollinator species depend on each other when surviving under difficult conditions.

Such a community-wide collapse, involving many pollinator species, can occur suddenly when increasingly harsh conditions pass a critical point and recovery from such a collapse might not be easy. The improvement in conditions needed for pollinators to recover, could be substantially larger than the improvement needed to return to conditions at which the pollinator community collapsed.

Pollen

Closeup image of a cactus flower and its stamens

Scanning electron microscope image of pollen grains from a variety of common plants: sunflower (*Helianthus annuus*), morning glory (*Ipomoea purpurea*), prairie hollyhock (*Sidalcea malviflora*), oriental lily (*Lilium auratum*), evening primrose (*Oenothera fruticosa*), and castor bean (*Ricinus communis*).

Pollen is a fine to coarse powdery substance comprising pollen grains which are male microgametophytes of seed plants, which produce male gametes (sperm cells). Pollen grains have a hard coat made of sporopollenin that protects the gametophytes during the process of their movement from the stamens to the pistil of flowering plants or from the male cone to the female cone of coniferous plants. If pollen lands on a compatible pistil or female cone, it germinates, producing a pollen tube that transfers the sperm to the ovule containing the female gametophyte. Individual pollen grains are small enough to require magnification to see detail. The study of pollen is called palynology and is highly useful in paleoecology, paleontology, archaeology, and forensics.

Pollen in plants is used for transferring haploid male genetic material from the anther of a single flower to the stigma of another in cross-pollination. In a case of self-pollination, this process takes place from the anther of a flower to the stigma of the same flower.

The Structure and Formation of Pollen

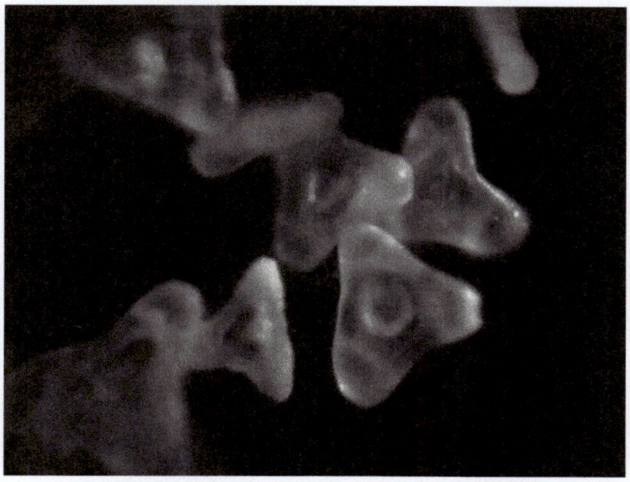

Triporate pollen of *Oenothera speciosa*

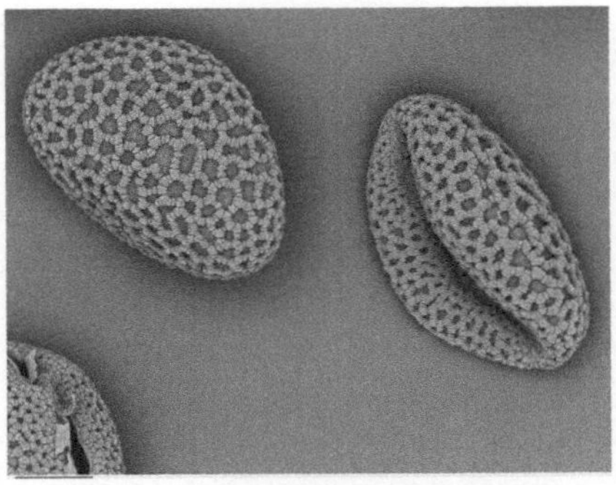

Pollen of *Lilium auratum* showing single sulcus (monosulcate)

Arabis pollen has three colpi and prominent surface structure.

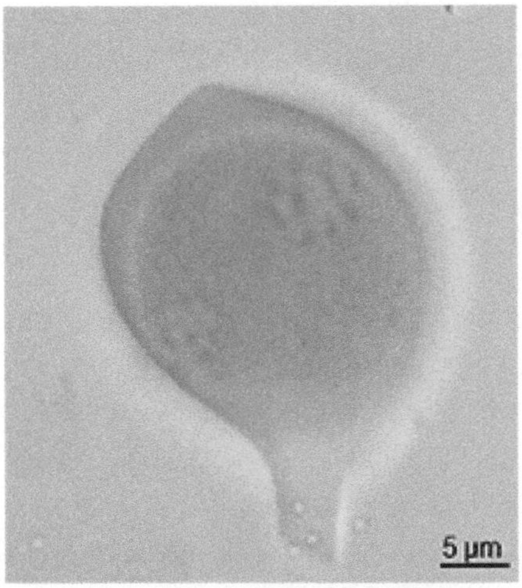

Pollens/Microspores of *Lycopersicon esculentum* at coenocytic tetrad stage of development observed through oil immersion microscope; the chromosomes of what will become four pollen grains can be seen.

Apple pollen under microscopy

Pollen itself is not the male gamete. Each pollen grain contains vegetative (non-reproductive) cells (only a single cell in most flowering plants but several in other seed plants) and a generative (reproductive) cell. In flowering plants the vegetative tube cell produces the pollen tube, and the generative cell divides to form the two sperm cells.

Formation

Pollen is produced in the microsporangia in the male cone of a conifer or other gymnosperm or in the anthers of an angiosperm flower. Pollen grains come in a wide variety of shapes, sizes, and surface markings characteristic of the species. Pollen grains of pines, firs, and spruces are winged. The smallest pollen grain, that of the forget-me-not (*Myosotis* spp.),is around 6 µm (0.006 mm) in diameter. Wind-borne pollen grains can be as large as about 90–100 µm.

In angiosperms, during flower development the anther is composed of a mass of cells that appear undifferentiated, except for a partially differentiated dermis. As the flower develops, four groups of sporogenous cells form within the anther. The fertile sporogenous cells are surrounded by layers of sterile cells that grow into the wall of the pollen sac. Some of the cells grow into nutritive cells that supply nutrition for the microspores that form by meiotic division from the sporogenous cells.

In a process called microsporogenesis, four haploid microspores are produced from each diploid sporogenous cell (microsporocyte, pollen mother cell or meiocyte), after meiotic division. After the formation of the four microspores, which are contained by callose walls, the development of the pollen grain walls begins. The callose wall is broken down by an enzyme called callase and the freed pollen grains grow in size and develop their characteristic shape and form a resistant outer wall called the exine and an inner wall called the intine. The exine is what is preserved in the fossil record. Two basic types of microsporogenesis are recognised, simultaneous and successive. In simultaneous microsporogenesis meiotic steps I and II are completed prior to cytokinesis, whereas in successive microsporogenesis cytokinesis follows. While there may be a continuum with intermediate forms, the type of microsporogenesis has systematic significance. The predominant form amongst the monocots is successive, but there are important exceptions.

During microgametogenesis, the unicellular microspores undergo mitosis and develop into mature microgametophytes containing the gametes. In some flowering plants, germination of the pollen grain may begin even before it leaves the microsporangium, with the generative cell forming the two sperm cells.

Structure

Except in the case of some submerged aquatic plants, the mature pollen grain has a double wall. The vegetative and generative cells are surrounded by a thin delicate wall of unaltered cellulose called the endospore or intine, and a tough resistant outer cuticularized wall composed largely of sporopollenin called the exospore or exine. The exine often bears spines or warts, or is variously sculptured, and the character of the markings is often of value for identifying genus, species, or even cultivar or individual. The spines may be less than a micron in length (spinulus, plural spinuli) referred to as spinulose (scabrate), or longer than a micron (echina, echinae) referred to as echinate. Various terms also describe the sculpturing such as reticulate, a net like appearance consisting of elements (murus, muri) separated from each other by a lumen (plural lumina).

The pollen wall protects the sperm while the pollen grain is moving from the anther to the stigma; it protects the vital genetic material from drying out and solar radiation. The pollen grain surface is covered with waxes and proteins, which are held in place by structures called sculpture elements on the surface of the grain. The outer pollen wall, which prevents the pollen grain from shrinking and crushing the genetic material during desiccation, is composed of two layers. These two layers are the tectum and the foot layer, which is just above the intine. The tectum and foot layer are separated by a region called the columella, which is composed of strengthening rods. The outer wall is constructed with a resistant biopolymer called sporopollenin.

Pollen apertures are regions of the pollen wall that may involve exine thinning or a significant reduction in exine thickness. They allow shrinking and swelling of the grain caused by changes in moisture content. Elongated apertures or furrows in the pollen grain are called colpi (singular: colpus) or sulci (singular: sulcus). Apertures that are more circular are called pores. Colpi, sulci and pores are major features in the identification of classes of pollen. Pollen may be referred to as inaperturate (apertures absent) or aperturate (apertures present). The aperture may have a lid (operculum), hence is described as operculate. However the term inaperturate covers a wide range of morphological types, such as functionally inaperturate (cryptoaperturate) and omniaperturate. Inaperaturate pollen grains often have thin walls, which facilitates pollen tube germination at any position. Terms such as uniaperturate and triaperturate refer to the number of apertures present (one and three respectively).

The orientation of furrows (relative to the original tetrad of microspores) classifies the pollen as sulcate or colpate. Sulcate pollen has a furrow across the middle of what was the outer face when the pollen grain was in its tetrad. If the pollen has only a single sulcus, it is described as monosulcate, has two sulci, as bisulcate, or more, as polysulcate. Colpate pollen has furrows other than across the middle of the outer faces. Eudicots have pollen with three colpi (tricolpate) or with shapes that are evolutionarily derived from tricolpate pollen. The evolutionary trend in plants has been from monosulcate to polycolpate or polyporate pollen.

Pollination

European honey bee carrying pollen in a pollen basket back to the hive

Marmalade hoverfly, pollen on its face and legs, sitting on a rockrose.

Diadasia bee straddles flower carpels while visiting yellow *Opuntia engelmannii* cactus

The transfer of pollen grains to the female reproductive structure (*pistil* in angiosperms) is called pollination. This transfer can be mediated by the wind, in which case the plant is described as **an**emophilous (literally wind-loving). Anemophilous plants typically produce great quantities of very lightweight pollen grains, sometimes with air-sacs. Non-flowering seed plants (e.g. pine trees) are characteristically anemophilous. Anemophilous flowering plants generally have inconspicuous flowers. Entomophilous (literally insect-loving) plants produce pollen that is relatively heavy, sticky and protein-rich, for dispersal by insect pollinators attracted to their flowers. Many insects and some mites are specialized to feed on pollen, and are called palynivores.

In non-flowering seed plants, pollen germinates in the pollen chamber, located beneath the micropyle, underneath the integuments of the ovule. A pollen tube is produced, which grows into the nucellus to provide nutrients for the developing sperm cells. Sperm cells of Pinophyta and Gnetophyta are without flagella, and are carried by the pollen tube, while those of Cycadophyta and Ginkgophyta have many flagella.

When placed on the stigma of a flowering plant, under favorable circumstances, a pollen grain puts forth a pollen tube, which grows down the tissue of the style to the ovary, and makes its way along the placenta, guided by projections or hairs, to the micropyle of an ovule. The nucleus of the tube cell has meanwhile passed into the tube, as does also the generative nucleus, which divides (if it hasn't already) to form two sperm cells. The sperm cells are carried to their destination in the tip of the pollen tube.

Pollen in the Fossil Record

Pollen's sporopollenin outer sheath affords it some resistance to the rigours of the fossilisation process that destroy weaker objects; it is also produced in huge quantities. There is an extensive fossil record of pollen grains, often disassociated from their parent plant. The discipline of palynology is devoted to the study of pollen, which can be used both for biostratigraphy and to gain information about the abundance and variety of plants alive — which can itself yield important information about paleoclimates. Pollen is first found in the fossil record in the late Devonian period and increases in abundance until the present day.

Allergy to Pollen

Nasal allergy to pollen is called pollinosis, and allergy specifically to grass pollen is called hay fever. Generally, pollens that cause allergies are those of anemophilous plants (pollen is dispersed by air currents.) Such plants produce large quantities of lightweight pollen (because wind dispersal is random and the likelihood of one pollen grain landing on another flower is small), which can be carried for great distances and are easily inhaled, bringing it into contact with the sensitive nasal passages.

In the US, people often mistakenly blame the conspicuous goldenrod flower for allergies. Since this plant is entomophilous (its pollen is dispersed by animals), its heavy, sticky pollen does not become independently airborne. Most late summer and fall pollen allergies are probably caused by ragweed, a widespread anemophilous plant.

Arizona was once regarded as a haven for people with pollen allergies, although several ragweed species grow in the desert. However, as suburbs grew and people began establishing irrigated lawns and gardens, more irritating species of ragweed gained a foothold and Arizona lost its claim of freedom from hay fever.

Anemophilous spring blooming plants such as oak, birch, hickory, pecan, and early summer grasses may also induce pollen allergies. Most cultivated plants with showy flowers are entomophilous and do not cause pollen allergies.

The number of people in the United States affected by hay fever is between 20 and 40 million, and such allergy has proven to be the most frequent allergic response in the nation. There are certain evidential suggestions pointing out hay fever and similar allergies to be of hereditary origin. Individuals who suffer from eczema or are asthmatic tend to be more susceptible to developing long-term hay fever.

In Denmark, decades of rising temperatures cause pollen to appear earlier and in greater numbers, as well as introduction of new species such as ragweed.

The most efficient way to handle a pollen allergy is by preventing contact with the material. Individuals carrying the ailment may at first believe that they have a simple summer cold, but hay fever becomes more evident when the apparent cold does not disappear. The confirmation of hay fever can be obtained after examination by a general physician.

Treatment

Antihistamines are effective at treating mild cases of pollinosis, this type of non-prescribed drugs includes loratadine, cetirizine and chlorpheniramine. They do not prevent the discharge of histamine, but it has been proven that they do prevent a part of the chain reaction activated by this biogenic amine, which considerably lowers hay fever symptoms.

Decongestants can be administered in different ways such as tablets and nasal sprays.

Allergy immunotherapy (AIT) treatment involves administering doses of allergens to accustom the body to pollen, thereby inducing specific long-term tolerance. Allergy immunotherapy can be administered orally (as sublingual tablets or sublingual drops), or by injections under the skin (subcutaneous). Discovered by Leonard Noon and John Freeman in 1911, allergy immunotherapy represents the only causative treatment for respiratory allergies.

Nutrition

Most major classes of predatory and parasitic arthropods contain species that eat pollen, despite the common perception that bees are the primary pollen-consuming arthropod group. Many other Hymenoptera other than bees consume pollen as adults, though only a small number feed on pollen as larvae (including some ant larvae). Spiders are normally considered carnivores but pollen is an important source of food for several species, particularly for spiderlings, which catch pollen on their webs. It is not clear how spiderlings manage to eat pollen however, since their mouths are not large enough to consume pollen grains. Some predatory mites also feed on pollen, with some species being able to subsist solely on pollen, such as *Euseius tularensis*, which feeds on the pollen of dozens of plant species. Members of some beetle families such as Mordellidae and Melyridae feed almost exclusively on pollen as adults, while various lineages within larger families such as Curculionidae, Chrysomelidae, Cerambycidae, and Scarabaeidae are pollen specialists even though most members of their families are not (e.g., only 36 of 40000 species of ground beetles, which are typically predatory, have been shown to eat pollen—but this is thought to be a severe underestimate as the feeding habits are only known for 1000 species). Similarly, Ladybird beetles mainly eat insects, but many species also eat pollen, as either part or all of their diet. Hemiptera are mostly herbivores or omnivores but pollen feeding is known (and has only been well studied in the Anthocoridae). Many adult flies, especially Syrphidae, feed on pollen, and three UK syrphid species feed strictly on pollen (syrphids, like all flies, cannot eat pollen directly due to the structure of their mouthparts, but can consume pollen contents that are dissolved in a fluid). Some species of fungus, including *Fomes fomentarius*, are able to break down grains of pollen as a secondary nutrition source that is particularly high in nitrogen. Pollen may be valuable diet supplement for detritivores, providing them with nutrients needed for growth, development and maturation. It was suggested that obtaining nutrients from pollen, deposited on the forest floor during periods of pollen rains, allows fungi to decompose nutritionally scarce litter.

Some species of *Heliconius* butterflies consume pollen as adults, which appears to be a valuable nutrient source, and these species are more distasteful to predators than the non-pollen consuming species.

Although bats, butterflies and hummingbirds are not pollen eaters *per se*, their consumption of nectar in flowers is an important aspect of the pollination process.

In Humans

A variety of producers have started selling bee pollen for human consumption, often marketed as a food (rather than a dietary supplement). The largest constituent is carbohydrates, with protein content ranging from 7 to 35 percent depending on the plant species collected by bees.

Honey produced by bees from natural sources contains pollen derived p-coumaric acid, an antioxidant.

The U.S. Food and Drug Administration (FDA) has not found any harmful effects of bee pollen consumption, except from the usual allergies. However, FDA does not allow bee pollen marketers in the United States to make health claims about their produce, as no scientific basis for these has ever been proven. Furthermore, there are possible dangers not only from allergic reactions but also from contaminants such as pesticides and from fungi and bacteria growth related to poor storage procedures. A manufacturers's claim that pollen collecting helps the bee colonies is also controversial.

Pine pollen is traditionally consumed in Korea as an ingredient in sweets and beverages.

Parasites

The growing industries in pollen harvesting for human and bee consumption rely on harvesting pollen baskets from honey bees as they return to their hives using a *pollen trap*. When this pollen has been tested for parasites, it has been found that a multitude of pollinator viruses and eukaryotic parasites are present in the pollen. It is currently unclear if the parasites are introduced by the bee that collected the pollen or if it is from contamination to the flower. Though this is not likely to pose a risk to humans, it is a major issue for the bumblebee rearing industry that relies on thousands of tonnes of honey bee collected pollen per year. Several sterilization methods have been employed, though no method has been 100% effective at sterilizing, without reducing the nutritional value, of the pollen.

Forensic Palynology

An SEM micrograph of Redbud pollen. Scanning electron microscopes are major instruments in palynology.

In forensic biology, pollen can tell a lot about where a person or object has been, because regions of the world, or even more particular locations such a certain set of bushes, will have a distinctive collection of pollen species. Pollen evidence can also reveal the season in which a particular object picked up the pollen. Pollen has been used to trace activity at mass graves in Bosnia, catch a burglar who brushed against a *Hypericum* bush during a crime, and has even been proposed as an additive for bullets to enable tracking them.

Pollen Tube

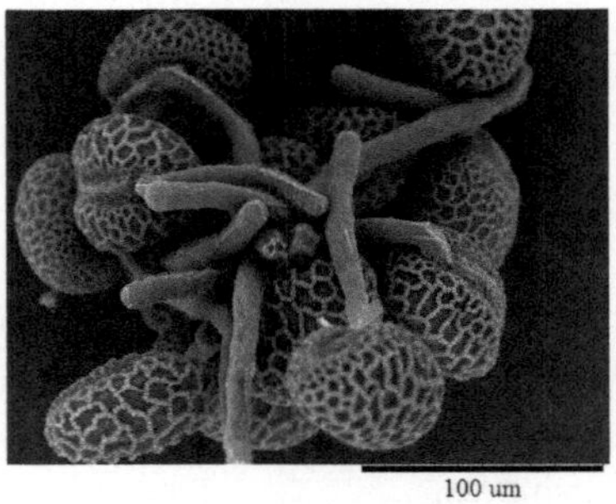

SEM image of pollen tubes growing from *Lily* pollen grains.

A pollen tube is part of the male gametophyte of seed plants. It acts as a conduit to transport the male gamete cells from the pollen grain, either from the stigma (in flowering plants) to the ovules at the base of the pistil, or directly through ovule tissue in some gymnosperms. In maize, this single cell can grow longer than 12 inches to traverse the length of the pistil.

Pollen tubes were first discovered by Giovanni Battista Amici.

The Pollen Tube in Angiosperms

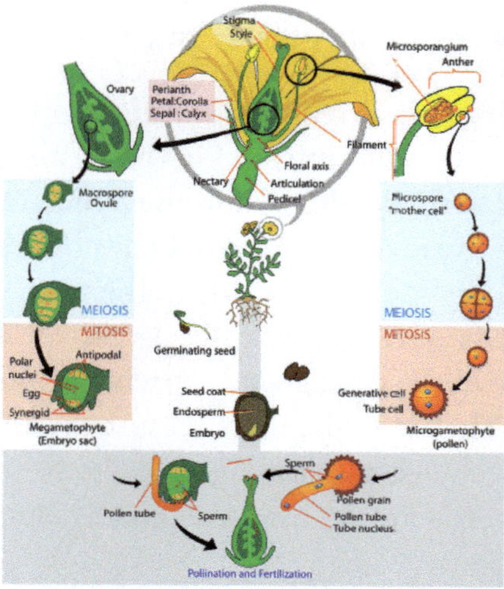

Angiosperm life cycle.

Angiosperm reproduction is a complex process that includes several steps that may vary among species. Each step is a vast procedure in its own right. Pollen is produced by the stamen, the male reproductive organ of the flower. Each pollen grain contains a vegetative cell, and a generative cell that divides to form two sperm cells. The pollen is delivered by the opening of anthers for subsequent pollination, that is, for the transfer of pollen grains to the pistil, the female reproductive organ. Pollination is usually carried out by wind, water or insects. The ovaries hold the ovules that produce the female gamete: the egg cell, which waits in place for fertilization.

Once a pollen grain settles on a compatible pistil, it may germinate in response to a sugary fluid secreted by the mature stigma of certain plants. Lipids at the surface of the stigma may also stimulate pollen tube growth for compatible pollen. Plants that are self-sterile often inhibit the pollen grains from their own flowers from growing pollen tubes. The presence of multiple grains of pollen has been observed to stimulate quicker pollen tube growth in some plants. The vegetative cell then produces the pollen tube, a tubular protrusion from the pollen grain, which carries the sperm cells within its cytoplasm. The sperm cells are the male gametes that will join with the egg cell and the central cell in double fertilization.

The germinated pollen tube must then drill its way through the nutrient-rich style and curl to the bottom of the ovary to reach the ovule. Once the pollen tube successfully attains an ovule, it delivers the two sperm cells with a burst. One of them fertilizes the egg cell to form an embryo, which will become the future plant. The other one fuses with both polar nuclei of the central cell to form the endosperm, which serves as the embryo's food supply. Finally, the ovary will develop into a fruit and the ovules will develop into seeds.

Pollen Tubes Used to Study Cell Physiology

Pollen tubes are an excellent model for the understanding of plant cell behavior. They are easily cultivated in vitro and have a very dynamic cytoskeleton that polymerizes at very high rates, providing the pollen tube with interesting mechanical properties. The pollen tube has an unusual kind of growth; it extends exclusively at its apex. Extending the cell wall only at the tip minimizes friction between the tube and the invaded tissue. This tip growth is performed in a pulsating manner rather than in a steady fashion. Remarkably, the pollen tube's journey through the style often results in depth-to-diameter ratios above 100:1 and up to 1000:1 in certain species. The internal machinery and the external interactions that govern the dynamics of pollen tube growth are far from being fully understood.

Pollen Tube Guidance

Extensive work has been dedicated to comprehend how the pollen tube responds to extracellular guidance signals to achieve fertilization. It is believed that pollen tubes react to a combination of chemical, electrical, and mechanical cues during its journey through the pistil. However, it is not clear how these external cues work or how they are processed internally. Moreover, sensory receptors for any external cue have not been identified yet. Nevertheless, several aspects have already been identified as central in the process of pollen tube growth. The actin filaments in the cytoskeleton, the peculiar cell wall, secretory vesicle dynamics, and the flux of ions, to name a few, are some of the fundamental features readily identified as crucial, but whose role has not yet been completely elucidated.

Self-pollination

One type of automatic self-pollination occurs in the orchid *Ophrys apifera*.
One of the two pollinia bends itself towards the stigma.

Self-pollination is when pollen from the same plant arrives at the stigma of a flower (in flowering plants) or at the ovule (in Gymnosperms). There are two types of self-pollination: In autogamy, pollen is transferred to the stigma of the same flower. In geitonogamy, pollen is transferred from the anther of one flower to the stigma of another flower on the same flowering plant, or from microsporangium to ovule within a single (monoecious) Gymnosperm. Some plants have mechanisms that ensure autogamy, such as flowers that do not open (cleistogamy), or stamens that move to come into contact with the stigma. The term selfing that is often used as a synonym, is not limited to self-pollination, but also applies to other types of (self fertilizations).

Occurrence

Few plants self-pollinate without the aid of pollen vectors (such as wind or insects). The mechanism is seen most often in some legumes such as peanuts. In another legume, soybeans, the flowers open and remain receptive to insect cross pollination during the day. If this is not accomplished, the flowers self-pollinate as they are closing. Among other plants that can self-pollinate are many kinds of orchids, peas, sunflowers and tridax. Most of the self-pollinating plants have small, relatively inconspicuous flowers that shed pollen directly onto the stigma, sometimes even before the bud opens. Self-pollinated plants expend less energy in the production of pollinator attractants and can grow in areas where the kinds of insects or other animals that might visit them are absent or very scarce—as in the Arctic or at high elevations.

Self-pollination limits the variety of progeny and may depress plant vigor. However, self-pollination can be advantageous, allowing plants to spread beyond the range of suitable pollinators or produce offspring in areas where pollinator populations have been greatly reduced or are naturally variable.

Pollination can also be accomplished by cross-pollination. Cross-pollination is the transfer of pollen, by wind or animals such as insects and birds, from the anther to the stigma of flowers on separate plants.

Types of Flowers that Self Pollinate

Both hermaphrodite and monoecious species have the potential for self-pollination leading to self-fertilization unless there is a mechanism to avoid it. Eighty percent of all flowering plants are hermaphroditic, meaning they contain both sexes in the same flower, while 5 percent of plant species are monoecious. The remaining 15% would therefore be dioecious (each plant unisexual).

Advantages of Self-pollination

There are several advantages for self-pollinating flowers. If a given genotype is well-suited for an environment, self-pollination helps to keep this trait stable in the species. Not being dependent on pollinating agents allows self-pollination to occur when bees and wind are nowhere to be found. Self-pollination can be an advantage when the number of flowers are small or widely spaced. During self-pollination the pollen grains are not transmitted from one flower to another. As a result the wastage of pollen is less. Also it does not depend on any external carrier. The plant that develops in that way of pollination cannot make changes in their characters and so the features of a species can be maintained. This way a species can maintain purity in it.

Disadvantages of Self-pollination

The disadvantages of self-pollination come from a lack of variation that allows no adaptation to the changing environment or potential pathogen attack. Self-pollination can lead to inbreeding depression caused by expression of deleterious recessive mutations, or to the reduced health of the species, due to the breeding of related specimens. This is why many flowers that could potentially self-pollinate have a built-in mechanism to avoid it, or make it second choice at best. Genetic defects in self-pollinating plants cannot be eliminated by genetic recombination and offspring can only avoid inheriting the deleterious attributes through a chance mutation arising in a gamete.

Mixed Mating

About 42% of flowering plants exhibit a mixed mating system in nature. In the most common kind of system, individual plants produce a single flower type and fruits may contain self-pollinated, out-crossed or a mixture of progeny types. Another mixed mating system is referred to as dimorphic cleistogamy. In this system a single plant produces both open, potentially out-crossed and closed, obligately self-pollinated cleistogamous flowers.

Self-pollinating Species

The evolutionary shift from outcrossing to self-fertilization is one of the most common evolutionary transitions in plants. About 10-15% of flowering plants are predominantly self-fertilizing. A few well-studied examples of self-pollinating species are described below.

Paphiopedilum Parishii

Self-pollination in the slipper orchid *Paphiopedilum parishii* occurs when the anther changes from a solid to a liquid state and directly contacts the stigma surface without the aid of any pollinating agent.

Holcoglossum Amesianum

The tree-living orchid *Holcoglossum amesianum* has a type of self-pollination mechanism in which the bisexual flower turns its anther against gravity through 360° in order to insert pollen into its own stigma cavity---without the aid of any pollinating agent or medium. This type of self-pollination appears to be an adaptation to the windless, drought conditions that are present when flowering occurs, at a time when insects are scarce. Without pollinators for outcrossing, the necessity of ensuring reproductive success appears to outweigh potential adverse effects of inbreeding. Such an adaptation may be widespread among species in similar environments.

Caulokaempferia Coenobialis

In the Chinese herb *Caulokaempferia coenobialis* a film of pollen is transported from the anther (pollen sacs) by an oily emulsion that slides sideways along the flower's style and into the individual's own stigma. The lateral flow of the film of pollen along the style appears to be due solely to the spreading properties of the oily emulsion and not to gravity. This strategy may have evolved to cope with a scarcity of pollinators in the extremely shady and humid habitats of *C. coenobialis*.

Capsella Rubella

Capsella rubella (Red Shepard's purse) is a self-pollinating species that became self-compatible 50,000 to 100,000 years ago, indicating that self-pollination is an evolutionary adaptation that can persist over many generations. Its out-crossing progenitor was identified as *Capsella grandiflora*.

Arabidopsis Thaliana

Arabidopsis thaliana is a predominantly self-pollinating plant with an out-crossing rate in the wild estimated at less than 0.3%. A study suggested that self-pollination evolved roughly a million years ago or more.

Bulbophyllum Bicoloratum

Self-pollination in the Madagascan orchid Bulbophyllum bicoloratum occurs by virtue of a rostellum that may have regained its stigmatic function as part of the distal median stigmatic lobe.

Possible Long-term Benefit of Meiosis

Meiosis followed by self-pollination produces little overall genetic variation. This raises the question of how meiosis in self-pollinating plants is adaptively maintained over extended periods [i.e. for roughly a million years or more, as in the case of *A.thaliana*] in preference to a less complicated and less costly asexual ameiotic process for producing progeny. An adaptive benefit of meiosis that may explain its long-term maintenance in self-pollinating plants is efficient recombinational repair of DNA damage. This benefit can be realized at each generation (even when genetic variation is not produced).

Autogamy

Autogamy, or self-fertilization, refers to the fusion of two gametes that come from one individual. Autogamy is predominantly observed in the form of self-pollination, a reproductive mechanism employed by many flowering plants. However, species of protists have also been observed using autogamy as a means of reproduction. Flowering plants engage in autogamy regularly, while the protists that engage in autogamy only do so in stressful environments.

Occurrence

Paramecium Aurelia

Paramecium aurelia is the most commonly studied protozoan for autogamy. Similar to other unicellular organisms, *Paramecium aurelia* typically reproduce asexually via binary fission or sexually via cross-fertilization. However, studies have shown that when put under nutritional stress, *Paramecium aurelia* will undergo meiosis and subsequent fusion of gametic-like nuclei. This process, defined as hemixis, a chromosomal rearrangement process, takes place in a number of steps. First, the two micronuclei of Paramecium aurelia enlarge and divide two times to form eight nuclei. Some of these daughter nuclei will continue to divide to create potential future gametic nuclei. Of these potential gametic nuclei, one will divide two more times. Of the four daughter nuclei arising from this step, two of them become anlagen, or cells that will form part of the new organism. The other two daughter nuclei become the gametic micronuclei that will undergo autogamous self-fertilization. These nuclear divisions are observed mainly when the Paramecium aurelia is put under nutritional stress. Research shows that *Paramecium aurelia* undergo autogamy synchronously with other individuals of the same species.

Tetrahymena Rostrata

Similar to Paramecium aurelia, the parasitic ciliate *Tetrahymena rostrata* has also been shown to engage in nuclear division and autogamy when placed under nutritional stress. Due to the degeneration and remodeling of genetic information that occurs in autogamy, genetic variability arises and possibly increases an offspring's chances of survival in stressful environments.

Allogromia Laticollaris

Allogromia laticollaris is perhaps the best-studied foraminiferan amoeboid for autogamy. *Allogromia laticollaris* can alternate between sexual reproduction via cross-fertilization and asexual reproduction via binary fission. The details of the life cycle of Allogromia laticollaris are unknown, but similar to *Paramecium aurelia, Allogromia laticollaris* is also shown to sometimes defer to autogamous behavior when placed in nutritional stress. As seen in Paramecium, there is some nuclear dimorphism observed in *Allogromia laticollaris.* There are often observations of macronuclei and chromosomal fragments coexisting in *Allogromia laticollaris.* This is indicative of nuclear and chromosomal degeneration, a process similar to the subdivisions observed in *Paramecium aurelia.* Multiple generations of haploid *Allogromia laticollaris* individuals can exist before autogamy actually takes place. The autogamous behavior in Allogromia laticollaris has the added consequence of giving rise to daughter cells that are substantially smaller than those rising from binary fission. It is hypothesized that this is a survival mechanism employed when the cell is in stressful environments, and thus not able to allocate all resources to creating offspring. If a cell was under nutritional stress and not able to function regularly, there would be a strong possibility of its offspring's fitness being sub-par.

Self-pollination in Flowering Plants

Self-pollination is an example of autogamy that occurs in flowering plants. Self-pollination occurs when the sperm in the pollen from the stamen of a plant goes to the carpels of that same plant and fertilizes the egg cell present. Self-pollination can either be done completely autogamously or geitonogamously. In the former, the egg and sperm cells that united came from the same flower. In the latter, the sperm and egg cells can come from a different flower on the same plant. While the latter method does blur the lines between autogamous self-fertilization and normal sexual reproduction, it is still considered autogamous self-fertilization.

Advantages of Autogamy

There are several advantages for the self-fertilization observed in flowering plants and protists. In flowering plants, it is important for some plants not to be dependent on pollinating agents that other plants rely for fertilization. This is unusual, however, considering that many plant species have evolved to become incompatible with their own gametes. While these species would not be well served by having autogamous self-fertilization as a reproductive mechanism, other species, which do not have self-incompatibility, would benefit from autogamy. Protists have the advantage of diversifying their modes of reproduction. This is useful for a multitude of reasons. First, if there is an unfavorable change in the environment that puts the ability to deliver offspring at risk, then it is advantageous for an organism to have autogamy at its disposal. In other organisms, it is seen that genetic diversity arising from sexual reproduction is maintained by changes in the environment that favor certain genotypes over others. Aside from extreme circumstances, it is possible that this form of reproduction gives rise to a genotype in the offspring that will increase fitness in the environment. This is due to the nature of the genetic degeneration and remodeling intrinsic to autogamy in unicellular organisms. Thus, autogamous behavior may become advantageous to have if an individual wanted to ensure offspring viability and survival. This advantage also applies to flowering plants. However, it is important to note that this change has not shown to produce a progeny with more fitness in unicellular organisms. It is possible that the nutrition deprived state of the parent cells before autogamy created a barrier for producing offspring that could thrive in those same stressful environments.

Disadvantages of Autogamy

In flowering plants, autogamy has the disadvantage of producing low genetic diversity in the species that use it as the predominant mode of reproduction. This leaves those species particularly susceptible to pathogens and viruses that can harm it. In addition, the foraminiferans that use autogamy have shown to produce substantially smaller progeny as a result. This indicates that since it is generally an emergency survival mechanism for unicellular species, the mechanism does not have the nutritional resources that would be provided by the organism if it were undergoing binary fission.

Evolution of Autogamy

Since autogamy in flowering plants and autogamy in unicellular species is fundamentally different, and plants and protists are not related, it is likely that both instances evolved separately. In flowering plants, it is believed that autogamy evolved one million years ago. However, due to the little overall genetic variation that arises in progeny, it is not fully understood how autogamy has been maintained in the tree of life.

Geitonogamy

Geitonogamy (from Greek *geiton* = neighbor + *gamein* = to marry) is a type of self-pollination. Geitonogamous pollination is sometimes distinguished from the fertilizations that can result from it, geitonogamy. If a plant is self-incompatible, geitonogamy can reduce seed production.

In flowering plants, pollen is transferred from a flower to another flower on the same plant, and in animal pollinated systems this is accomplished by a pollinator visiting multiple flowers on the same plant. Geitonogamy is also possible within species that are wind-pollinated, and may actually be a quite common source of self-fertilized seeds in self-compatible species. It also occurs in monoecious gymnosperms. Although geitonogamy is functionally cross-pollination involving a pollinating agent, genetically it is similar to autogamy since the pollen grains come from the same plant.

Monoecious plants like maize show geitonogamy. Geitonogamy is not possible for strictly dioecious plants.

Xenogamy

Xenogamy (Greek *xenos*=stranger, *gamos*=marriage) is the transfer of pollen grains from the anther to the stigma of a different plant. This is the only type of pollination which during pollination brings genetically different types of pollen grains to the stigma.

The term xenogamy (along with geitonogamy and autogamy) was first suggested by Kerner in 1876. Cross-pollination involves the transfer of pollen grains from the flower of one plant to the stigma of the flower of another plant.

The main characteristics which facilitate cross-pollination are:

- Herkogamy: Flowers possess some mechanical barrier on their stigmatic surface to avoid self-pollination, e.g. presence of gynostegium and pollinia in *Calotropis*.
- Dichogamy: Pollen and stigma of the flower mature at different times to avoid self-pollination.
- Self-incompatibility: In same plants, the mature pollen fall on the receptive stigma of the same flower but fail to bring about self-pollination.
- Male sterility: The pollen grains of some plants are not functional. Such plants set seeds only after cross-pollination.
- Dioecism: Cross-pollination always occurs when the plants are unisexual and dioecious, i.e., male and female flowers occur on separate plants, e.g., papaya, some cucurbits, etc.
- Heterostyly: The flowers of some plants have different lengths of stamens and styles so that self-pollination is not possible, e.g., *Primula*, *Linum*, etc.

Self-incompatibility in Plants

Self-incompatibility (SI) is a general name for several genetic mechanisms in angiosperms, which prevent self-fertilization and thus encourage outcrossing and allogamy. It should not be confused with genetically controlled physical or temporal mechanisms that prevent self-pollination, such as heterostyly and sequential hermaphroditism (dichogamy).

In plants with SI, when a pollen grain produced in a plant reaches a stigma of the same plant or another plant with a similar genotype, the process of pollen germination, pollen-tube growth, ovule fertilization and embryo development is halted at one of its stages and consequently no seeds are produced. SI is one of the most important means of preventing inbreeding and promoting the generation of new genotypes in plants, and it is considered as one of the causes for the spread and success of angiosperms on the earth.

Mechanisms of Single-locus Self-incompatibility

The best studied mechanisms of SI act by inhibiting the germination of pollen on stigmas, or the elongation of the pollen tube in the styles. These mechanisms are based on protein-protein interactions, and the best-understood mechanisms are controlled by a single locus termed S, which has many different alleles in the species population. Despite their similar morphological and genetic manifestations, these mechanisms have evolved independently, and are based on different cellular components; therefore, each mechanism has its own, unique S-genes.

The S-locus contains two basic protein coding regions - one expressed in the pistil, and the other in the anther and/or pollen (referred to as the female and male determinants, respectively). Because of their physical proximity, these are genetically linked, and are inherited as a unit. The units are called S-haplotypes. The translation products of the two regions of the S-locus are two proteins which, by interacting with one another, lead to the arrest of pollen germination and/or pollen tube elongation, and thereby generate an SI response, preventing fertilization. However, when a female determinant interacts with a male determinant of a different haplotype, no SI is created, and fertilization ensues. This is a simplistic description of the general mechanism of SI, which is more complicated, and in some species the S-haplotype contains more than two protein coding regions.

Following is a detailed description of the different known mechanisms of SI in plants.

Gametophytic Self-incompatibility (GSI)

In gametophytic self-incompatibility (GSI), the SI phenotype of the pollen is determined by its own gametophytic haploid genotype. This is the more common type of SI. Two different mechanisms of GSI have been described in detail at the molecular level, and their description follows.

The RNase Mechanism

The female component of GSI in the Solanaceae was found in 1989. Proteins in the same family were subsequently discovered in the Rosaceae and Plantaginaceae. Despite some early doubts about the common ancestry of GSI in these distantly related families, phylogenetic studies and the finding of shared male determinants (F-box proteins) clearly established homology. Consequently, this mechanism arose approximately 90 million years ago, and is the inferred ancestral state for approximately 50% of all plants.

In this mechanism, pollen tube elongation is halted when it has proceeded approximately one third of the way through the style. The female component ribonuclease, termed S-RNase probably causes degradation of the ribosomal RNA (rRNA) inside the pollen tube, in the case of identical male and female S alleles, and consequently pollen tube elongation is arrested, and the pollen grain dies.

The male component was only recently putatively identified as a member of the "F-box" protein family. Despite some fairly convincing evidence that it may be the male component, several features also make it an unlikely candidate.

The S-glycoprotein Mechanism

The following mechanism was described in detail in *Papaver rhoeas*. In this mechanism, pollen growth is inhibited within minutes of its placement on the stigma.

The female determinant is a small, extracellular molecule, expressed in the stigma; the identity of the male determinant remains elusive, but it is probably some cell membrane receptor. The interaction between male and female determinants transmits a cellular signal into the pollen tube, resulting in strong influx of calcium cations; this interferes with the intracellular concentration gradient of calcium ions which exists inside the pollen tube, essential for its elongation. The influx of calcium ions arrests tube elongation within 1–2 minutes. At this stage, pollen inhibition is still reversible, and elongation can be resumed by applying certain manipulations, resulting in ovule fertilization.

Subsequently, the cytosolic protein p26, a pyrophosphatase, is inhibited by phosphorylation, possibly resulting in arrest of synthesis of molecular building blocks, required for tube elongation. There is depolymerization and reorganization of actin filaments, within the pollen cytoskeleton. Within 10 minutes from the placement on the stigma, the pollen is committed to a process which ends in its death. At 3–4 hours past pollination, fragmentation of pollen DNA begins, and finally (at 10–14 hours), the cell dies apoptotically.

Sporophytic Self-incompatibility (SSI)

In sporophytic self-incompatibility (SSI), the SI phenotype of the pollen is determined by the diploid genotype of the anther (the sporophyte) in which it was created. This form of SI was identified in the families: Brassicaceae, Asteraceae, Convolvulaceae, Betulaceae, Caryophyllaceae, Sterculiaceae and Polemoniaceae. Up to this day, only one mechanism of SSI has been described in detail at the molecular level, in *Brassica* (Brassicaceae).

Since SSI is determined by a diploid genotype, the pollen and pistil each express the translation products of two different alleles, i.e. two male and two female determinants. Dominance relationships often exist between pairs of alleles, resulting in complicated patterns of compatibility/self-incompatibility. These dominance relationships also allow the generation of individuals homozygous for a recessive S allele.

Compared to a population in which all S alleles are co-dominant, the presence of dominance relationships in the population, raises the chances of compatible mating between individuals. The frequency ratio between recessive and dominant S alleles, reflects a dynamic balance between reproduction assurance (favoured by recessive alleles) and avoidance of selfing (favoured by dominant alleles).

The SI Mechanism in *Brassica*

As previously mentioned, the SI phenotype of the pollen is determined by the diploid genotype of the anther. In *Brassica*, the pollen coat, derived from the anther's tapetum tissue, carries the translation products of the two S alleles. These are small, cysteine-rich proteins. The male determinant is termed SCR or SP11, and is expressed in the anther tapetum as well as in the microspore and pollen (i.e. sporophytically). There are possibly up to 100 polymorphs of the S-haplotype in Brassica, and within these there is a dominance hierarchy.

The female determinant of the SI response in *Brassica*, is a transmembrane protein termed SRK, which has an intracellular kinase domain, and a variable extracellular domain. SRK is expressed in the stigma, and probably functions as a receptor for the SCR/SP11 protein in the pollen coat. Another stigmatic protein, termed SLG, is highly similar in sequence to the SRK protein, and seems to function as a co-receptor for the male determinant, amplifying the SI response.

The interaction between the SRK and SCR/SP11 proteins results in autophosphorylation of the intracellular kinase domain of SRK, and a signal is transmitted into the papilla cell of the stigma. Another protein essential for the SI response is MLPK, a serine-threonine kinase, which is anchored to the plasma membrane from its intracellular side. The downstream cellular and molecular events, leading eventually to pollen inhibition, are poorly described.

Other Mechanisms of self-incompatibility

These mechanisms have received only limited attention in scientific research. Therefore, they are still poorly understood.

2-locus Gametophytic Self-incompatibility

The grass subfamily Pooideae, and perhaps all of the family Poaceae, have a gametophytic self-incompatibility system that involves two unlinked loci referred to as *S* and *Z*. If the alleles expressed at these two loci in the pollen grain both match the corresponding alleles in the pistil, the pollen grain will be recognized as incompatible.

Heteromorphic Self-incompatibility

A distinct SI mechanism exists in heterostylous flowers, termed heteromorphic self-incompatibility. This mechanism is probably not evolutionarily related to the more familiar mechanisms, which are differentially defined as homomorphic self-incompatibility.

Almost all heterostylous taxa feature SI to some extent. The loci responsible for SI in heterostylous flowers, are strongly linked to the loci responsible for flower polymorphism, and these traits are inherited together. Distyly is determined by a single locus, which has two alleles; tristyly is determined by two loci, each with two alleles. Heteromorphic SI is sporophytic, i.e. both alleles in the male plant, determine the SI response in the pollen. SI loci always contain only two alleles in the population, one of which is dominant over the other, in both pollen and pistil. Variance in SI alleles parallels the variance in flower morphs, thus pollen from one morph can fertilize only pistils from the other morph. In tristylous flowers, each flower contains two types of stamens; each stamen produces pollen capable of fertilizing only one flower morph, out of the three existing morphs.

A population of a distylous plant contains only two SI genotypes: ss and Ss. Fertilization is possible only between genotypes; each genotype cannot fertilize itself. This restriction maintains a 1:1 ratio between the two genotypes in the population; genotypes are usually randomly scattered in space. Tristylous plants contain, in addition to the S locus, the M locus, also with two alleles. The number of possible genotypes is greater here, but a 1:1 ratio exists between individuals of each SI type.

Cryptic Self-incompatibility (CSI)

Cryptic self-incompatibility (CSI) exists in a limited number of taxa (for example, there is evidence for CSI in *Silene vulgaris*, Caryophyllaceae). In this mechanism, the simultaneous presence of cross and self pollen on the same stigma, results in higher seed set from cross pollen, relative to self pollen. However, as opposed to 'complete' or 'absolute' SI, in CSI, self-pollination without the presence of competing cross pollen, results in successive fertilization and seed set; in this way, reproduction is assured, even in the absence of cross-pollination. CSI acts, at least in some species, at the stage of pollen tube elongation, and leads to faster elongation of cross pollen tubes, relative to self pollen tubes. The cellular and molecular mechanisms of CSI have not been described.

The strength of a CSI response can be defined, as the ratio of crossed to selfed ovules, formed when equal amounts of cross and self pollen, are placed upon the stigma; in the taxa described up to this day, this ratio ranges between 3.2 and 11.5.

Late-acting Self-incompatibility (LSI)

Late-acting self-incompatibility (LSI) is also termed ovarian self-incompatibility (OSI). In this mechanism, self pollen germinates and reaches the ovules, but no fruit is set. LSI can be pre-zygotic (e.g. deterioration of the embryo sac prior to pollen tube entry, as in *Narcissus triandrus*) or post-zygotic (malformation of the zygote or embryo, as in certain species of *Asclepias* and in *Spathodea campanulata*).

The existence of the LSI mechanism among different taxa and in general, is subject for scientific debate. Criticizers claim, that absence of fruit set is due to genetic defects (homozygosity for lethal recessive alleles), which are the direct result of self-fertilization (inbreeding depression). Supporters, on the other hand, argue for the existence of several basic criteria, which differentiate certain cases of LSI from the inbreeding depression phenomenon.

Self-compatibility (SC)

Approximately one half of angiosperm species are SI, the remainder being self-compatible (SC). Mutations that break down SI (resulting in SC) may become common or entirely dominate in natural populations. Pollinator decline, variability in pollinator service, the so-called "automatic advantage" of self-fertilisation, among other factors, may favor the loss of SI. Similarly, human-mediated artificial selection through selective breeding may be responsible for the commonly observed SC in cultivated plants. SC enables more efficient breeding techniques to be employed for crop improvement.

Fruit Tree Pollination

A European honey bee pollinates a peach flower while collecting nectar.

Pollination of fruit trees is required to produce seeds with surrounding fruit. It is the process of moving pollen from the anther to the stigma, either in the same flower or in another flower. Some tree species, including many fruit trees, do not produce fruit from self-pollination, so pollinizer trees are planted in orchards.

The pollination process requires a carrier for the pollen, which can be animal, wind, or human intervention (by hand-pollination or by using a pollen sprayer). Cross pollination produces seeds with a different genetic makeup from the parent plants; such seeds may be created deliberately as part of a selective breeding program for fruit trees with desired attributes. Trees that are cross-pollinated or pollinated via an insect pollinator produce more fruit than trees with flowers that just self-pollinate. In fruit trees, bees are an essential part of the pollination process for the formation of fruit.

Pollination of fruit trees around the world has been highly studied for hundreds of years. There is a lot of information known about fruit tree pollination from temperate climates, but very little information known about fruit tree pollination from tropical climates. Fruits from temperate climates include apples, pears, plums, peaches, cherries, berries, grapes, and nuts which are considered dry fruits. Fruits from tropical climates include bananas, pineapples, papayas, passion fruit, avocado, mango, and members of the genus *Citrus*.

Apple

Most apples are self-incompatible, that is, they do not produce fruit when pollinated from a flower of the same tree or from another tree of the same cultivar, and must be cross pollinated. A few are described as "self-fertile" and are capable of self-pollination, although even those tend to carry larger crops when cross pollinated from a suitable pollenizer. A relatively small number of cultivars are "triploid", meaning that they provide almost no viable pollen for themselves or other apple trees. Apples that can pollinate one another are grouped by the time they flower so cross-pollinators are in bloom at the same time. Pollination management is an important component of apple culture. Before planting, it is important to arrange for pollenizers - varieties of apple or crabapple that provide plentiful, viable and compatible pollen. Orchard blocks may alternate rows of compatible varieties, or may plant crabapple trees, or graft on limbs of crabapple. Some varieties produce very little pollen, or the pollen is sterile, so these are not good pollenizers. Good-quality nurseries have pollenizer compatibility lists. Growers with old orchard blocks of single varieties sometimes provide bouquets of crabapple blossoms in drums or pails in the orchard for pollenizers. Home growers with a single tree and no other variety in the neighborhood can do the same on a smaller scale.

During the bloom each season, commercial apple growers usually provide pollinators to carry the pollen. Honeybee hives are most commonly used in the United States, and arrangements may be made with a commercial beekeeper who supply hives for a fee. Honeybees of the genus *Apis* are the most common pollinator for apple trees, although members of the genera *Andrena, Bombus, Halictus*, and *Osmia* pollinate apple trees in the wild. Solitary bees such as ground-nesting mining bees (*Andrena*) may play a far bigger role in pollination than at one time suspected. Bumble bees are sometimes present in orchards, but not usually in enough quantity to be significant pollinators; in the home garden with only a few trees, their role may be much greater.

Increasingly Orchard bees (spring mason bees) are being used in fruit tree pollination. According to British writer Christopher O'Toole in his book *The Red Mason Bee*, *Osmia rufa* is a much more efficient pollinator of orchard crops (in Europe) than honey bees. Both *O. rufa* and *O. cornuta* are used in Europe, while in western North America, the "Blue Orchard Bee" (*Osmia lignaria*, more black than blue in color) is a proven orchard pollinator. In Japan, the Japanese Orchard Bee—the hornfaced bee, *Osmia cornifrons*—provides up to 80% of the apple pollination. Beyond Japan, *Osmia cornifrons* is also used increasingly in the eastern US, because like other mason bees it is up to 100 times more efficient than the honeybee—a mere 600 hornfaced bees being required per hectare, as opposed to tens of thousands of honeybees. Home growers may find these more acceptable in suburban locations because they rarely sting.

Symptoms of inadequate pollination are small and misshapen apples, and slowness to ripen. The seeds can be counted to evaluate pollination. Well-pollinated apples have best quality, and will have seven to ten seeds. Apples with fewer than three seeds will usually not mature and will drop from the trees in the early summer. Inadequate pollination can result from either a lack of pollinators or pollenizers, or from poor pollinating weather at bloom time. Multiple bee visits are usually required to deliver sufficient grains of pollen to accomplish complete pollination.

Almonds

The blossoms of most California Almond varieties are self-incompatible, requiring cross-pollination with other varieties to produce a crop. The single most important factor determining a good yield is pollination during the bloom period. More than a million colonies of honey bees are placed in California Almond orchards at the beginning of the bloom period to pollinate the crop. California beekeepers alone cannot supply this critical need, which is why honey bees travel across the country to the San Joaqin Valley each year. Although the recommended number of hives per acre is 2 to 3, due to the high demand in conjunction with the reduced availability of commercial beehives, many almond growers have to make do with a lower hive density during pollination. These growers started using semiochemical formulations, like SPLAT Bloom, to compensate for the low hive density. SPLAT Bloom manipulates the behavior of the bees, inciting them to spend more time foraging, and thus pollinating flowers in entire the almond orchard (increasing pollination and fruit set), not only close to the hive.

Research into self-fertile almonds has led to the development of several almond varieties that do not need a pollinator tree. Working from an old Spanish variety, "Tuono", researchers in 2008 started making available new self-fertile varieties that more closely match the qualities of the popular "Nonpareil" almond—including varieties "Lone Star" and "Independence". While self-fertile trees do not require pollen from a second tree variety for fruit / nut set, they still depend upon insect pollination for good production. Almond growers with self-fertile almonds report excellent nut set with half (or less) the number of bees in the field.

Pear

Like apples, pears are self-incompatible and need to attract insects in order to be pollinated and produce fruit. One notable difference from apples is that pear blossoms are much less attractive to bees due to their pale coloring and light odor. Bees may abandon the pear blossoms to visit dandelions or a nearby apple orchard. The majority of pollinators of pear trees are honey bees, although pears are also visited by blow flies and hoverflies. A way to combat the low attraction of honey bees to pear blossoms is to use bee attractants to entice the bees to pollinate the flowers. Bee attractants may include pheromones that mimic the brood pheromone or the juvenile pheromone, or other attractants. There are also other methods for attracting honey bees to pear blossoms. One is saturation pollination, that is to stock so many bees that all area blossoms are worked regardless of the attractiveness to the bees. Another method is to delay the movement of the beehives into the orchards until there is about 30 percent bloom. The bees are moved into the orchard during the night and will usually visit the pear blossoms for a few hours until they discover the richer nectar sources. The recommended number of hives per acre is one.

Citrus

Many citrus cultivars are seedless and are produced parthenocarpically without pollination. Some cultivars may be capable of producing fruit either way, having seeds in the segments, if pollinated, and no seeds if not. Citrus that requires pollination may be self-compatible, thus pollen must be moved only a short distance from the anther to the stigma by a pollinator. Some citrus, such as Meyer Lemons, are popular container plants. When these bloom indoors, they often suffer from blossom drop because no pollinators have access. Hand pollination by a human pollinator is a solution.

A few citrus, including some tangelos and tangerines, are self-incompatible, and require cross pollination. Pollinizers must be planned when groves are planted. Managed honeybee hives at bloom time are often used to ensure adequate pollination.

Pollinator

A syrphid fly (*Eristalinus taeniops*) pollinating a common hawkweed
A mining bee (*Andrena lonicerae*) pollinating a honeysuckle (*Lonicera gracilipes*).

A pollinator is an animal that moves pollen from the male anther of a flower to the female stigma of a flower. This helps to bring about fertilization of the ovules in the flower by the male gametes from the pollen grains.

Insect pollinators include bees, (honey bees, solitary species, bumblebees); pollen wasps (Masarinae); ants; flies including bee flies and hoverflies; lepidopterans, both butterflies and moths; and flower beetles. Vertebrates, mainly bats and birds, but also some non-bat mammals (monkeys, lemurs, possums, rodents) and some lizards pollinate certain plants. Among the pollinating birds are hummingbirds, honeyeaters and sunbirds with long beaks; they pollinate a number of deep-throated flowers.

Cycads, which are not flowering plants, are also pollinated by insects. A pollinator is different from a pollenizer, a plant that is a source of pollen for the pollination process.

Background

Plants fall into pollination syndromes that reflect the type of pollinator being attracted. These are characteristics such as: overall flower size, the depth and width of the corolla, the color (including patterns called nectar guides that are visible only in ultraviolet light), the scent, amount of nectar, composition of nectar, etc. For example, birds visit red flowers with long, narrow tubes and lots of nectar, but are not as strongly attracted to wide flowers with little nectar and copious pollen, which are more attractive to beetles. When these characteristics are experimentally modified (altering colour, size, orientation), pollinator visitation may decline.

It has recently been discovered that cycads, which are not flowering plants, are also pollinated by insects.

Types of Pollinators

Bees

Lipotriches sp. bee pollinating flowers

An Australian painted lady with its proboscis extended during feeding. Butterflies are recognised pollinators though not as effective as bees.

The most recognized pollinators are the various species of bees, which are plainly adapted to pollination. Bees typically are fuzzy and carry an electrostatic charge. Both features help pollen grains adhere to their bodies, but they also have specialized pollen-carrying structures; in most bees, this takes the form of a structure known as the scopa, which is on the hind legs of most bees, and/or the lower abdomen (e.g., of megachilid bees), made up of thick, plumose setae. Honey bees, bumblebees, and their relatives do not have a scopa, but the hind leg is modified into a structure called the corbicula (also known as the "pollen basket"). Most bees gather nectar, a concentrated energy source, and pollen, which is high protein food, to nurture their young, and inadvertently transfer some among the flowers as they are working. Euglossine bees pollinate orchids, but these are male bees collecting floral scents rather than females gathering nectar or pollen. Female orchid bees act as pollinators, but of flowers other than orchids. Eusocial bees such as honey bees need an abundant and steady pollen source to multiply.

Honey Bees

Honey bees travel from flower to flower, collecting nectar (later converted to honey), and pollen grains. The bee collects the pollen by rubbing against the anthers. The pollen collects on the hind legs, in a structure referred to as a "pollen basket". As the bee flies from flower to flower, some of the pollen grains are transferred onto the stigma of other flowers.

Nectar provides the energy for bee nutrition; pollen provides the protein. When bees are rearing large quantities of brood (beekeepers say hives are "building"), bees deliberately gather pollen to meet the nutritional needs of the brood.

Good pollination management seeks to have bees in a "building" state during the bloom period of the crop, thus requiring them to gather pollen, and making them more efficient pollinators. Thus, the management techniques of a beekeeper providing pollination services are different from, and to some extent in tension with, those of a beekeeper who is trying to produce honey.

Millions of hives of honey bees are contracted out as pollinators by beekeepers, and honey bees are by far the most important commercial pollinating agents, but many other kinds of pollinators, from blue bottle flies, to bumblebees, orchard mason bees, and leaf cutter bees are cultured and sold for managed pollination.

Other species of bees differ in various details of their behavior and pollen-gathering habits, and honey bees are not native to the Western Hemisphere; all pollination of native plants in the Americas historically has been performed by various native bees.

Other Insects

Many insects other than bees accomplish pollination by visiting flowers for nectar or pollen, or commonly both. Many do so adventitiously, but the most important pollinators are specialists for at least parts of their lifecycles for at least certain functions. For example, males of many species of Hymenoptera, including many hunting wasps, rely on freely flowering plants as sources of energy (in the form of nectar) and also as territories for meeting fertile females that visit the flowers. Prominent examples are predatory wasps (especially Sphecidae, Vespidae, and Pompilidae). The term "pollen wasps", in particular, is widely applied to the Masarinae, a subfamily of the Vespidae;

they are remarkable among solitary wasps in that they specialise in gathering pollen for feeding their larvae, carried internally and regurgitated into a mud chamber prior to oviposition.

Many bee flies, and some Tabanidae and Nemestrinidae are particularly adapted to pollinating fynbos and Karoo plants with narrow, deep corolla tubes, such as *Lapeirousia* species. Part of the adaptation takes the form of remarkably long probosces.

Lepidoptera (butterflies and moths) also pollinate plants to various degrees. They are not major pollinators of food crops, but various moths are important pollinators of other commercial crops such as tobacco. Pollination by certain moths may be important, however, or even crucial, for some wildflowers mutually adapted to specialist pollinators. Spectacular examples include orchids such as *Angraecum sesquipedale*, dependant on a particular hawk moth, Morgan's sphinx. Yucca species provide other examples, being fertilised in elaborate ecological interactions with particular species of yucca moths.

Beetles of species that specialise in eating pollen, nectar, or flowers themselves, are important cross-pollinators of some plants such as members of the Araceae and Zamiaceae, that produce prodigious amounts of pollen. Others, for example the Hopliini, specialise in free-flowering species of the Asteraceae and Aizoaceae.

Various midges and thrips are comparatively minor opportunist pollinators. Ants also pollinate some kinds of flowers, but for the most part they are parasites, robbing nectar without conveying useful amounts of pollen to a stigma. Whole groups of plants, such as certain fynbos *Moraea* and *Erica* species produce flowers on sticky peduncles or with sticky corolla tubes that only permit access to flying pollinators, whether bird, bat, or insect.

Carrion flies and flesh flies in families such as Calliphoridae and Sarcophagidae are important for some species of plants whose flowers exude a fetid odor. The plants' ecological strategy varies; several species of *Stapelia*, for example, attract carrion flies that futilely lay their eggs on the flower, where their larvae promptly starve for lack of carrion. Other species do decay rapidly after ripening, and offer the visiting insects large masses of food, as well as pollen and sometimes seed to carry off when they leave.

Hoverflies are important pollinators of flowering plants worldwide. Often hoverflies are considered to be the second most important pollinators after wild bees. Although hoverflies as a whole are generally considered to be nonselective pollinators, some species have more specialized relationships. The orchid species *Epipactis veratrifolia* mimics alarm pheromones of aphids to attract hover flies for pollination. Another plant, the slipper orchid in southwest China, also achieves pollination by deceit by exploiting the innate yellow colour preference of syrphide.

Some male *Bactrocera* fruit flies are exclusive pollinators of some wild *Bulbophyllum* orchids that lack nectar and have a specific chemical attractant and reward (methyl eugenol, raspberry ketone or zingerone) present in their floral fragrances.

A class of strategy of great biological interest is that of sexual deception, where plants, generally orchids, produce remarkably complex combinations of pheromonal attractants and physical mimicry that induce male bees or wasps to attempt to mate with them, conveying pollinia in the process. Examples are known from all continents apart from Antarctica, though Australia appears to be exceptionally rich in examples.

Some Diptera (flies) may be the main pollinators at higher elevations of mountains, whereas *Bombus* species are the only pollinators among Apoidea in alpine regions at timberline and beyond.

Other insect orders are rarely pollinators, and then typically only incidentally (e.g., Hemiptera such as Anthocoridae and Miridae).

Vertebrates

Bats are important pollinators of some tropical flowers. Birds, particularly hummingbirds, honeyeaters and sunbirds also accomplish much pollination, especially of deep-throated flowers. Other vertebrates, such as kinkajous, monkeys, lemurs, possums, rodents and lizards have been recorded pollinating some plants.

Humans can be pollinators, as many gardeners have discovered that they must hand pollinate garden vegetables, whether because of pollinator decline (as has been occurring in parts of the U.S. since the mid-20th century) or simply to keep a strain genetically pure. This can involve using a small brush or cotton swab to move pollen, or to simply tap or shake tomato blossoms to release the pollen for the self pollinating flowers. Tomato blossoms are self-fertile, but (with the exception of potato-leaf varieties) have the pollen inside the anther, and the flower requires shaking to release the pollen through pores. This can be done by wind, by humans, or by a sonicating bee (one that vibrates its wing muscles while perched on the flower), such as a bumblebee. Sonicating bees are extremely efficient pollinators of tomatoes, and colonies of bumblebees are quickly replacing humans as the primary pollinators for greenhouse tomatoes.

Pollinator Population Declines and Conservation

Pollinators provide a key ecosystem service vital to the maintenance of both wild and agricultural plant communities. In 1999 the Convention on Biological Diversity issued the São Paulo Declaration on Pollinators, recognizing the critical role that these species play in supporting and maintaining terrestrial productivity as well as the survival challenges they face due to anthropogenic change. Today pollinators are considered to be in a state of decline; some species, such as Franklin's bumble bee (*Bombus franklini*) have been red-listed and are in danger of extinction. Although managed bee hives are increasing worldwide, these can not compensate for the loss of wild pollinators in many locations.

Declines in the health and population of pollinators pose what could be a significant threat to the integrity of biodiversity, to global food webs, and to human health. At least 80% of our world's crop species require pollination to set seed. An estimated one out of every three bites of food comes to us through the work of animal pollinators. The quality of pollinator service has declined over time and this had led to concerns that pollination will be less resistant to extinction in the future.

United States National Strategy to Promote the Health of Honey Bees and Other Pollinators

In recent times, environmental groups have put pressure on the Environmental Protection Agency to ban neonicotinoids, a type of insecticide. In May 2015, the Obama Administration released a strategy called National Strategy to Promote the Health of Honey Bees and Other Pollinators. The administration announced it would include input from the pesticide industry in putting together the initiative.

The task force goal is "tackling and reducing the impact of multiple stressors on pollinator health, including pests and pathogens, reduced habitat, lack of nutritional resources, and exposure to pesticides."

The EPA and U.S. Department of Agriculture are leading the task force.

The Structure of Plant-pollinator Networks

Wild pollinators often visit a large number of plant species and plants are visited by a large number of pollinator species. All these relations together form a network of interactions between plants and pollinators. Surprising similarities were found in the structure of networks consisting out of the interactions between plants and pollinators. This structure was found to be similar in very different ecosystems on different continents, consisting of entirely different species.

The structure of plant-pollinator networks may have large consequences for the way in which pollinator communities respond to increasingly harsh conditions. Mathematical models, examining the consequences of this network structure for the stability of pollinator communities suggest that the specific way in which plant-pollinator networks are organized minimizes competition between pollinators and may even lead to strong indirect facilitation between pollinators when conditions are harsh. This allows pollinator species to survive together under harsh conditions. But it also means that pollinator species collapse simultaneously when conditions pass a critical point. This simultaneous collapse occurs, because pollinator species depend on each other when surviving under difficult conditions.

Such a community-wide collapse, involving many pollinator species, can occur suddenly when increasingly harsh conditions pass a critical point and recovery from such a collapse might not be easy. The improvement in conditions needed for pollinators to recover, could be substantially larger than the improvement needed to return to conditions at which the pollinator community collapsed.

Pollination Syndrome

Baltimore (*Euphydryas phaeton*) nectaring at daisy (*Argyranthemum*)

Pollination syndromes are suites of flower traits that have evolved in response to natural selection imposed by different pollen vectors, which can be abiotic (wind and water) or biotic, such as birds, bees, flies, and so forth. These traits include flower shape, size, colour, odour, reward type and amount, nectar composition, timing of flowering, etc. For example, tubular red flowers with copious nectar often attract birds; foul smelling flowers attract carrion flies or beetles, etc.

The "classical" pollination syndromes as they are currently defined were developed in the 19th century by the Italian botanist Federico Delpino. Although they have been useful in developing our understanding of plant-pollinator interactions, an uncritical acceptance of pollination syndromes as providing a framework for classifying these relationships is rather out of date.

Abiotic

These do not attract animal pollinators. Nevertheless, they often have suites of shared traits.

Plantago media, pollinated by wind or insects

Wind Pollination (Anemophily)

Flowers may be small and inconspicuous, as well as green and not showy. They produce enormous numbers of relatively small pollen grains (hence wind-pollinated plants may be allergens, but seldom are animal-pollinated plants allergenic). Their stigmas may be large and feathery to catch the pollen grains. Insects may visit them to collect pollen; in some cases, these are ineffective pollinators and exert little natural selection on the flowers, but there are also examples of ambophilous flowers which are both wind and insect pollinated. Anemophilous, or wind pollinated flowers, are usually small and inconspicuous, and do not possess a scent or produce nectar. The anthers may produce a large number of pollen grains, while the stamens are generally long and protrude out of flower.

Water Pollination (Hydrophily)

Water-pollinated plants are aquatic and pollen is released into the water. Water currents therefore act as a pollen vector in a similar way to wind currents. Their flowers tend to be small and inconspicuous with lots of pollen grains and large, feathery stigmas to catch the pollen. However, this is relatively uncommon (only 2% of pollination is hydrophily) and most aquatic plants are insect-pollinated, with flowers that emerge into the air. *Vallisneria* is an example.

Biotic

Sunflower pollinated by butterflies and bees

Bee Pollination (Melittophily)

Bee-pollinated flowers can be very variable in their size, shape and colouration. They can be open and bowl-shaped (radially symmetrical) or more complex and non-radially symmetric ("zygomorphic"), as is the case with many peas and foxgloves.

Some bee flowers tend to be yellow or blue, often with ultraviolet nectar guides and scent. Nectar, pollen, or both are offered as rewards in varying amounts. The sugar in the nectar tends to be sucrose-dominated. A few bees collect oil from special glands on the flower.

There are diverse types of bees (such as honeybees, bumblebees, and orchid bees), forming large groups that are quite distinctive in size, tongue length and behaviour (some solitary, some colonial); thus generalization about bees is difficult. Some plants can only be pollinated by bees because their anthers release pollen internally, and it must be shaken out by buzz pollination (also known as "sonication"). Bees are the only animals that perform this behaviour. Bumblebees sonicate, but honeybees do not.

Wasp Pollination

Wasps are also responsible for the pollination of several plants species, being important pollen vectors, and in some cases, even more efficient pollinators than bees.

Butterfly Pollination (Psychophily)

Hesperoyucca whipplei (moth-pollinated)

Butterfly-pollinated flowers tend to be large and showy, pink or lavender in colour, frequently have a landing area, and are usually scented. Since butterflies do not digest pollen (with one exception), more nectar is offered than pollen. The flowers have simple nectar guides with the nectaries usually hidden in narrow tubes or spurs, reached by the long tongue of the butterflies.

Moth Pollination (Phalaenophily)

Day-flying sphinx moth nectaring on Brazilian vervain

Among the more important moth pollinators are the hawk moths (Sphingidae). Their behaviour is similar to hummingbirds: they hover in front of flowers with rapid wingbeats. Most are nocturnal or crepuscular. So moth-pollinated flowers tend to be white, night-opening, large and showy with tubular corollas and a strong, sweet scent produced in the evening, night or early morning. A lot of nectar is produced to fuel the high metabolic rates needed to power their flight.

Other moths (Noctuids, Geometrids, Pyralids, for example) fly slowly and settle on the flower. They do not require as much nectar as the fast-flying hawk moths, and the flowers tend to be small (though they may be aggregated in heads).

Sapromyophilous *Stapelia gigantea*

Fly Pollination (Myophily and Sapromyophily)

Flies tend to be important pollinators in high-altitude and high-latitude systems, where they are numerous and other insect groups may be lacking. There are two main types of fly pollination: myophily and sapromyophily.

Myophily includes flies that feed on nectar and pollen as adults - particularly bee flies (Bombyliidae), hoverflies (Syrphidae), and others - and these regularly visit flowers. In contrast, male fruit flies (Tephritidae) are enticed by specific floral attractants emitted by some wild orchids which do not produce nectar. Chemicals emitted by the orchid act as the fly's sex pheromone precursor or booster. Myophilous plants tend not to emit a strong scent, are typically purple, violet, blue, and white, and have open dishes or tubes.

Sapromyophiles, on the other hand, normally visit dead animals or dung. They are attracted to flowers which mimic the odor of such objects. The plant provides them with no reward and they leave quickly unless it has traps to slow them down. Such plants are far less common than myophilous ones.

Bird Pollination (Ornithophily)

Although hummingbirds are the most familiar nectar-feeding birds for North Americans, there are analogous species in other parts of the world: sunbirds, honeyeaters, flowerpeckers, honeycreepers, bananaquits, flowerpiercers, lories and lorikeets. Hummingbirds are the oldest group, with the greatest degree of specialization on nectar. Flowers attractive to hummingbirds that can hover in front of the flower tend to be large red or orange tubes with a lot of dilute nectar, secreted during the day. Since birds do not have a strong response to scent, they tend to be odorless. Perching birds need a substantial landing platform, so sunbirds, honeyeaters, and the like are less associated with tubular flowers.

African baobab (bat-pollinated)

Bat Pollination (Chiropterophily)

Bat-pollinated flowers tend to be large and showy, white or light coloured, open at night and have strong odours. They are often large and bell-shaped. Bats drink the nectar, and these plants typically offer nectar for extended periods of time. Sight, smell, and echo-location are used to initially find the flowers, and excellent spatial memory is used to visit them repeatedly. In fact, bats can identify nectar-producing flowers using echolocation. In the New World, bat pollinated flowers often have sulfur-scented compounds, but this does not carry to other parts of the world. Bat-pollinated plants have bigger pollen than their relatives.

Beetle Pollination (Cantharophily)

Beetle-pollinated flowers are usually large, greenish or off-white in color and heavily scented. Scents may be spicy, fruity, or similar to decaying organic material. Most beetle-pollinated flowers are flattened or dish shaped, with pollen easily accessible, although they may include traps to keep the beetle longer. The plant's ovaries are usually well protected from the biting mouthparts of their pollinators. Beetles may be particularly important in some parts of the world such as semi-arid areas of southern Africa and southern California and the montane grasslands of KwaZulu-Natal in South Africa.

Biology

Pollination syndromes reflect convergent evolution towards forms (phenotypes) that limit the number of species of pollinators visiting the plant. They increase the functional specialization of the plant with regard to pollination, though this may not affect the ecological specialization (i.e. the number of species of pollinators within that functional group). They are responses to common selection pressures exerted by shared pollinators or abiotic pollen vectors, which generate correlations among traits. That is, if two distantly related plant species are both pollinated by nocturnal moths, for example, their flowers will converge on a form which is recognised by the moths (e.g. pale colour, sweet scent, nectar released at the base of a long tube, night-flowering).

Advantages of Specialization

- Efficiency of pollination: the rewards given to pollinators (commonly nectar or pollen or both, but sometimes oil, scents, resins, or wax) may be costly to produce. Nectar can be cheap, but pollen is generally expensive as it is relatively high in nitrogen compounds. Plants have evolved to obtain the maximum pollen transfer for the minimum reward delivered. Different pollinators, because of their size, shape, or behaviour, have different efficiencies of transfer of pollen. And the floral traits affect efficiency of transfer: columbine flowers were experimentally altered and presented to hawkmoths, and flower orientation, shape, and colour were found to affect visitation rates or pollen removal.

- Pollinator constancy: to efficiently transfer pollen, it is best for the plant if the pollinator focuses on one species of plant, ignoring other species. Otherwise, pollen may be dropped uselessly on the stigmas of other species. Animals, of course, do not aim to pollinate, they aim to collect food as fast as they can. However, many pollinator species exhibit constancy, passing up available flowers to focus on one plant species. Why should animals specialize on a plant species, rather than move to the next flower of any species? Although pollinator constancy was recognized by Aristotle, the benefits to animals are not yet fully understood.

> The most common hypothesis is that pollinators must learn to handle particular types of flowers, and they have limited capacity to learn different types. They can only efficiently gather rewards from one type of flower.

These honeybees selectively visit flowers from only one species for a period of time, as can be seen by the colour of the pollen in their baskets:

Advantages of Generalization

Pollinators fluctuate in abundance and activity independently of their plants, and any one species may fail to pollinate a plant in a particular year. Thus a plant may be at an advantage if it attracts several species or types of pollinators, ensuring pollen transfer every year. Many species of plants have the back-up option of self-pollination, if they are not self-incompatible.

Criticisms of the Syndromes

Whilst it is clear that pollination syndromes can be observed in nature, there has been much debate amongst scientists as to how frequent they are and to what extent we can use the classical syndromes to classify plant-pollinator interactions. Although some species of plants are visited only by one type of animal (i.e. they are functionally specialized), many plant species are visited by very different pollinators. For example, a flower may be pollinated by bees, butterflies, and birds. Strict specialization of plants relying on one species of pollinator is relatively rare, probably because it can result in variable reproductive success across years as pollinator populations vary significantly. In such cases, plants should generalize on a wide range of pollinators, and such ecological generalization is frequently found in nature. A study in Tasmania found the syndromes did not usefully predict the pollinators.

This debate has led to a critical re-evaluation of the syndromes, which suggests that on average about one third of the flowering plants can be classified into the classical syndromes. This reflects the fact that nature is much less predictable and straightforward than 19th Century biologists originally thought. Pollination syndromes can be thought of as extremes of a continuum of greater or lesser specialization or generalization onto particular functional groups of pollinators that exert similar selective pressures" and the frequency with which flowers conform to the expectations of the pollination syndromes is relatively rare. In addition, new types of plant-pollinator interaction, involving "unusual" pollinating animals are regularly being discovered, such as specialized pollination by spider hunting wasps (Pompilidae) and fruit chafers (Cetoniidae) in the eastern grasslands of South Africa. These plants do not fit into the classical syndromes, though they may show evidence of convergent evolution in their own right.

An analysis of flower traits and visitation in 49 species in the plant genus *Penstemon* found that it was possible to separate bird- and bee- pollinated species quite well, but only by using floral traits which were not considered in the classical accounts of syndromes, such as the details of anther opening. Although a recent review concluded that there is "overwhelming evidence that functional groups exert different selection pressures on floral traits", the sheer complexity and subtlety of plant-pollinator interactions (and the growing recognition that non-pollinating organisms such as seed predators can affect the evolution of flower traits) means that this debate is likely to continue for some time.

Hand-pollination

Hand pollination (also called "mechanical pollination") is a technique used when natural, or open pollination is insufficient or undesirable. The most common techniques are for crops such as cucurbits, which may exhibit poor pollination by fruit abortion, fruit deformity or poor maturation. Hand-pollination is an option only on a small scale, but is a common technique by home and small market gardeners, involving the transfer of pollen with an artist's brush or cotton swab from male to female flowers. Sometimes the corolla is removed from male flowers and the flower itself is brushed against the stigmas of female flowers.

The principal reasons for hand pollination include a paucity of natural pollinators, avoiding cross-pollination between varieties grown together or, conversely, in the controlled production of hybrids.

Hand-pollination can be the simple act of shaking, with tomatoes and other self-fertile flowers. In general, tomatoes grown in greenhouses need aid in pollination, which has been done by hand or with an electric vibrator for a long time. Recently, it has been mostly supplanted by cultured bumblebees.

When pollination is needed on a large scale, such as for field crops, orchards, or commercial seed production, honeybees or other pollinators are cultured and provided for pollination management.

The primary reason for hand-pollination is lack of pollinators, which may be due to the act of moving crops from home areas (as in vanilla beans) without also taking the natural pollinator.

Pollinator decline or the concentrated pollination needs of monoculture may also be other factors in pollination shortage. When growing plants in urban environments lack of pollinators is often an issue as well.

Another reason for hand-pollination is in order to control the parents of the seed to be produced. For example, all F1 hybrids are the result of pollinating one specific variety with another specific variety, in order to produce a uniform crop.

For pest prevention with some high-value crops, row covers of mesh are used to prevent pest insects from access to the plants. However, this may necessitate hand-pollination when bloom starts.

Hand-pollination is used with date palms to avoid wasting a lot of space and energy growing sufficient male plants for adequate natural pollination.

Variable techniques for hand-pollination have arisen for many specialty crops.

Pears grown in Hanyuan County, China have been hand-pollinated since the 1980s in order to produce better fruit. The Jinhuali pear was promoted by China authorities but farmers quickly discovered that it can't be pollinated with other varieties that have different flowering times. They tried hand pollination and it worked well. In the same pear, lice infestation requires the use of many insecticide sprays; in this condition, local beekeepers refuse to lend beehives.

References

- Rodríguez-Gironés, Miguel A.; Santamaría, Luis (2004). "Why are so many bird flowers red?". PLoS Biology. 2 (10): e306. doi:10.1371/journal.pbio.0020350. PMC 521733. PMID 15486585
- Raghavan, Valayamghat (1997). Molecular Embryology of Flowering Plants. Cambridge University Press. pp. 210–216. ISBN 978-0-521-55246-2
- Kuehn, F. Coordinator. (2015). Farming for native bees. World Wide Web electronic publication. Retrieved from [1]. (Accessed: September 22, 2015)
- Malhó, Rui (1998). "Pollen tube guidance – the long and winding road". Sexual Plant Reproduction. 11 (5): 242–244. doi:10.1007/s004970050148
- Sühs, R.B.; Somavilla, A.; Putzke, J.; Köhler, A. (2009). "Pollen vector wasps (Hymenoptera, Vespidae) of Schinus terebinthifolius Raddi (Anacardiaceae)". Brazilian Journal of Biosciences. Santa Cruz do Sul, RS, Brazil. pp. 138–143
- Cronk, J. K.; Fennessy, M. Siobhan (2001). Wetland plants: biology and ecology. Boca Raton, Fla.: Lewis Publishers. p. 166. ISBN 1-56670-372-7
- Judd, Walter S. & Olmstead, Richard G. (2004). "A survey of tricolpate (eudicot) phylogenetic relationships". American Journal of Botany. 91 (10): 1627–1644. doi:10.3732/ajb.91.10.1627. PMID 21652313
- Glover, Beverly J. (2007). Understanding flowers and flowering: an integrated approach. Oxford University Press. p. 127. ISBN 0-19-856596-8
- Siewertsen, Bjarne. "Hård nyser for allergikere i varm fremtid" (English: Hard sneeze for allergic people in warm future) Danish Meteorological Institute, 18 April 2015. Retrieved: 19 April 2015
- Skoner, DP (July 2001). "Allergic rhinitis: definition, epidemiology, pathophysiology, detection, and diagnosis.". The Journal of allergy and clinical immunology. 108 (1 Suppl): S2–8. PMID 11449200
- Simpson, Michael G. (2011). "Palynology". Plant Systematics. Academic Press. pp. 453–464. ISBN 978-0-08-051404-8. Retrieved 6 January 2014

- Robinson, Kenneth (1985). "The Responses of Cells to Electrical Fields: A Review". The Journal of Cell Biology. 101. doi:10.1083/jcb.101.6.2023. PMC 2114002. PMID 3905820
- "The Pollen Feeders". Relationships of Natural Enemies and Non-Prey Foods. 7. 2009. pp. 87–11. doi:10.1007/978-1-4020-9235-0_6. ISBN 978-1-4020-9234-3
- "News - Communications - University of Canterbury - New Zealand". Comsdev.canterbury.ac.nz. 2012-03-23. Retrieved 2012-04-21
- Hessing, M.B. (1988). "Geitonogamous Pollination and Its Consequences in Gernium caespitosum". American Journal of Botany. 75 (9): 1324–1333. doi:10.2307/2444455. JSTOR 2444455
- Schwarze, Francis W. M. R.; Engels, Julia; Mattheck, Claus (2000). Fungal Strategies of Wood Decay in Trees. Springer. p. 61. ISBN 978-3-540-67205-0
- "National Strategy to Promote the Health of Honey Bees and Other Pollinators" (PDF). The White House. 19 May 2015. Retrieved 29 September 2015
- Ganders, F. R. (1979). "The biology of heterostyly". New Zealand Journal of Botany. 17: 607–635. doi:10.1080/0028825x.1979.10432574
- Abrol, Dharam (2015). Pollination Biology, Vol. 1. Switzerland: Springer International Publishing. ISBN 978-3-319-21084-1
- Abrol, Dharam P. (2012). "Non Bee Pollinators-Plant Interaction". Pollination Biology. Chapter 9: 265–310. doi:10.1007/978-94-007-1942-2_9
- P.J. Gullan & P.S. Cranston (2005). The Insects: An Outline of Entomology. Blackwell Publishing Ltd. p. 282. ISBN 1-4051-1113-5
- Chittka, L.; Thomson, J.D.; Waser, N.M. (1999). "Flower constancy, insect psychology, and plant evolution". Naturwissenschaften. 86: 361–177. doi:10.1007/s001140050636
- McLaughlin, Chris (2010). "5. Pure Pollination". The Complete Idiot's Guide to Heirloom Vegetables. Penguin. p. 56. ISBN 978-1-61564-052-2. Retrieved July 5, 2011
- Dressler, Robert L. (1968-03-01). "Pollination by Euglossine Bees". Evolution. 22 (1): 202–210. doi:10.2307/2406664. JSTOR 2406664
- Tan, K.H., Nishida R. and Toong, Y.C. (2002) Bulbophyllum cheiri's floral synomone lures fruit flies to perform pollination. Journal of Chemical Ecology 28:1161-1172
- Rai, Nagendra; Rai, Mathura (2006). Heterosis breeding in vegetable crops. New India Publishing. ISBN 978-81-89422-03-5. Retrieved July 5, 2011
- Cardinal, Sophie; Danforth, Bryan N. (2013). "Bees diversified in the age of eudicots". Proceedings of the Royal Society. 280: 20122686. doi:10.1098/rspb.2012.2686. PMC 3574388. PMID 23363629
- Tan, K.H.& R. Nishida. 2005. Synomone or Kairomone? - Bulbophyllum apertum (Orchidaceae) flower releases raspberry ketone to attract Bactrocera fruit flies. Journal of Chemical Ecology. 31(3): 509-519
- Sporne, Kenneth R. (1972). "Some Observations on the Evolution of Pollen Types in Dicotyledons". New Phytologist. 71 (1): 181–185. doi:10.1111/j.1469-8137.1972.tb04826.x
- Hingston, AB & PB Mcquillan (2000). "Are pollination syndromes useful predictors of floral visitors in Tasmania?". Australian Journal of Ecology. 25 (6): 600–609. doi:10.1046/j.1442-9993.2000.01059.x

4

An Introduction to Plant Breeding

Plant breeding is the act of changing certain traits of plants to generate preferred features. The techniques used in the process of plant breeding range from simply selecting desirable characteristics in plants to complex molecular techniques. Doubled haploidy, plant propagation and genetically modified crops are some of the topics related to plant breeding that have been discussed in the following section. The topics discussed in the chapter are of great importance to broaden the existing knowledge on plant reproduction.

Plant Breeding

The Yecoro wheat (right) cultivar is sensitive to salinity, plants resulting from a hybrid cross with cultivar W4910 (left) show greater tolerance to high salinity

Plant breeding is the art and science of changing the traits of plants in order to produce desired characteristics. Plant breeding can be accomplished through many different techniques ranging from simply selecting plants with desirable characteristics for propagation, to methods that make use of knowledge of genetics and chromosomes, to more complex molecular techniques.

Plant breeding has been practiced for thousands of years, since near the beginning of human civilization. It is practiced worldwide by individuals such as gardeners and farmers, or by professional plant breeders employed by organizations such as government institutions, universities, crop-specific industry associations or research centers.

International development nation agencies believe that breeding new crops is important for ensuring food security by developing new varieties that are higher-yielding, disease resistant, drought-resistant or regionally adapted to different environments and growing conditions.

History

Plant breeding started with sedentary agriculture and particularly the domestication of the first agricultural plants, a practice which is estimated to date back 9,000 to 11,000 years. Initially early farmers simply selected food plants with particular desirable characteristics, and employed these as progenitors for subsequent generations, resulting in an accumulation of valuable traits over time.

Gregor Mendel's experiments with plant hybridization led to his establishing laws of inheritance. Once this work became well known, it formed the basis of the new science of genetics, which stimulated research by many plant scientists dedicated to improving crop production through plant breeding.

Modern plant breeding is applied genetics, but its scientific basis is broader, covering molecular biology, cytology, systematics, physiology, pathology, entomology, chemistry, and statistics (biometrics). It has also developed its own technology.

Classical Plant Breeding

One major technique of plant breeding is selection, the process of selectively propagating plants with desirable characteristics and eliminating or "culling" those with less desirable characteristics.

Another technique is the deliberate interbreeding (crossing) of closely or distantly related individuals to produce new crop varieties or lines with desirable properties. Plants are crossbred to introduce traits/genes from one variety or line into a new genetic background. For example, a mildew-resistant pea may be crossed with a high-yielding but susceptible pea, the goal of the cross being to introduce mildew resistance without losing the high-yield characteristics. Progeny from the cross would then be crossed with the high-yielding parent to ensure that the progeny were most like the high-yielding parent, (backcrossing). The progeny from that cross would then be tested for yield (selection, as described above) and mildew resistance and high-yielding resistant plants would be further developed. Plants may also be crossed with themselves to produce inbred varieties for breeding. Pollinators may be excluded through the use of pollination bags.

Classical breeding relies largely on homologous recombination between chromosomes to generate genetic diversity. The classical plant breeder may also make use of a number of *in vitro* techniques such as protoplast fusion, embryo rescue or mutagenesis to generate diversity and produce hybrid plants that would not exist in nature.

Traits that breeders have tried to incorporate into crop plants include:

1. Improved quality, such as increased nutrition, improved flavor, or greater beauty
2. Increased yield of the crop
3. Increased tolerance of environmental pressures (salinity, extreme temperature, drought)
4. Resistance to viruses, fungi and bacteria
5. Increased tolerance to insect pests
6. Increased tolerance of herbicides
7. Longer storage period for the harvested crop

Before World War II

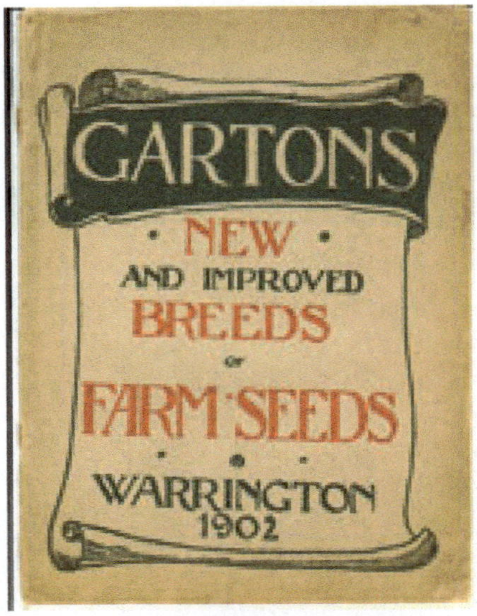

Garton's catalogue from 1902

Successful commercial plant breeding concerns were founded from the late 19th century. Gartons Agricultural Plant Breeders in England was established in the 1890s by John Garton, who was one of the first to commercialize new varieties of agricultural crops created through cross-pollination. The firm's first introduction was Abundance Oat, one of the first agricultural grain varieties bred from a *controlled* cross, introduced to commerce in 1892.

In the early 20th century, plant breeders realized that Mendel's findings on the non-random nature of inheritance could be applied to seedling populations produced through deliberate pollinations to predict the frequencies of different types. Wheat hybrids were bred to increase the crop production of Italy during the so-called "Battle for Grain" (1925–1940). Heterosis was explained by George Harrison Shull. It describes the tendency of the progeny of a specific cross to outperform both parents. The detection of the usefulness of heterosis for plant breeding has led to the development of inbred lines that reveal a heterotic yield advantage when they are crossed. Maize was the first species where heterosis was widely used to produce hybrids.

Statistical methods were also developed to analyze gene action and distinguish heritable variation from variation caused by environment. In 1933 another important breeding technique, cytoplasmic male sterility (CMS), developed in maize, was described by Marcus Morton Rhoades. CMS is a maternally inherited trait that makes the plant produce sterile pollen. This enables the production of hybrids without the need for labor-intensive detasseling.

These early breeding techniques resulted in large yield increase in the United States in the early 20th century. Similar yield increases were not produced elsewhere until after World War II, the Green Revolution increased crop production in the developing world in the 1960s.

After World War II

Following World War II a number of techniques were developed that allowed plant breeders to hybridize distantly related species, and artificially induce genetic diversity.

In vitro-culture of Vitis (grapevine), Geisenheim Grape Breeding Institute

When distantly related species are crossed, plant breeders make use of a number of plant tissue culture techniques to produce progeny from otherwise fruitless mating. Interspecific and intergeneric hybrids are produced from a cross of related species or genera that do not normally sexually reproduce with each other. These crosses are referred to as *Wide crosses*. For example, the cereal triticale is a wheat and rye hybrid. The cells in the plants derived from the first generation created from the cross contained an uneven number of chromosomes and as result was sterile. The cell division inhibitor colchicine was used to double the number of chromosomes in the cell and thus allow the production of a fertile line.

Failure to produce a hybrid may be due to pre- or post-fertilization incompatibility. If fertilization is possible between two species or genera, the hybrid embryo may abort before maturation. If this does occur the embryo resulting from an interspecific or intergeneric cross can sometimes be rescued and cultured to produce a whole plant. Such a method is referred to as Embryo Rescue. This technique has been used to produce new rice for Africa, an interspecific cross of Asian rice *(Oryza sativa)* and African rice *(Oryza glaberrima)*.

Hybrids may also be produced by a technique called protoplast fusion. In this case protoplasts are fused, usually in an electric field. Viable recombinants can be regenerated in culture.

Chemical mutagens like EMS and DMS, radiation and transposons are used to generate mutants with desirable traits to be bred with other cultivars - a process known as *Mutation Breeding*. Classical plant breeders also generate genetic diversity within a species by exploiting a process called somaclonal variation, which occurs in plants produced from tissue culture, particularly plants derived from callus. Induced polyploidy, and the addition or removal of chromosomes using a technique called chromosome engineering may also be used.

When a desirable trait has been bred into a species, a number of crosses to the favored parent are made to make the new plant as similar to the favored parent as possible. Returning to the example of the mildew resistant pea being crossed with a high-yielding but susceptible pea, to make the mildew resistant progeny of the cross most like the high-yielding parent, the progeny will be crossed back to that parent for several generations. This process removes most of the genetic contribution of the mildew resistant parent. Classical breeding is therefore a cyclical process.

With classical breeding techniques, the breeder does not know exactly what genes have been introduced to the new cultivars. Some scientists therefore argue that plants produced by classical breeding methods should undergo the same safety testing regime as genetically modified plants. There have been instances where plants bred using classical techniques have been unsuitable for human consumption, for example the poison solanine was unintentionally increased to unacceptable levels in certain varieties of potato through plant breeding. New potato varieties are often screened for solanine levels before reaching the marketplace.

Modern Plant Breeding

Modern plant breeding may use techniques of molecular biology to select, or in the case of genetic modification, to insert, desirable traits into plants. Application of biotechnology or molecular biology is also known as molecular breeding.

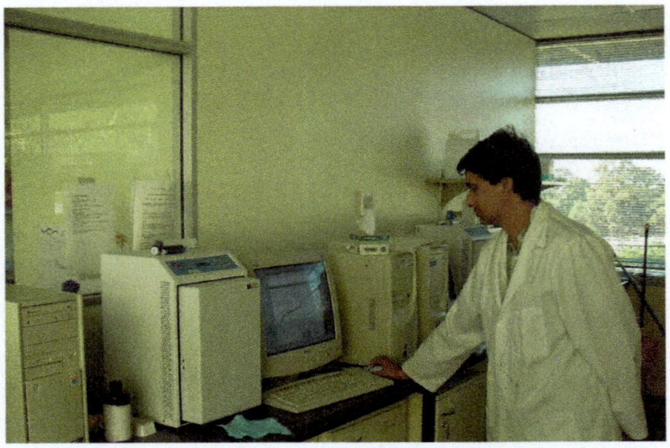

Modern facilities in molecular biology are now used in plant breeding.

Steps of Plant Breeding

The following are the major activities of plant breeding:

- Collection of variation
- Selection
- Evaluation
- Release
- Multiplication
- Distribution of the new variety
- Selling to people

Marker Assisted Selection

Sometimes many different genes can influence a desirable trait in plant breeding. The use of tools such as molecular markers or DNA fingerprinting can map thousands of genes. This allows plant breeders to screen large populations of plants for those that possess the trait of interest. The screening is based on the presence or absence of a certain gene as determined by laboratory procedures, rather than on the visual identification of the expressed trait in the plant.

Reverse Breeding and Doubled Haploids (DH)

Homozygous plants with desirable traits can be produced from heterozygous starting plants, if a haploid cell with the alleles for those traits can be produced, and then used to make a doubled haploid. The doubled haploid will be homozygous for the desired traits. Furthermore, two different homozygous plants created in that way can be used to produce a generation of F1 hybrid plants which have the advantages of heterozygosity and a greater range of possible traits. Thus, an individual heterozygous plant chosen for its desirable characteristics can be converted into a heterozygous variety (F1 hybrid) without the necessity of vegetative reproduction but as the result of the cross of two homozygous/doubled haploid lines derived from the originally selected plant.

Genetic Modification

Genetic modification of plants is achieved by adding a specific gene or genes to a plant, or by knocking down a gene with RNAi, to produce a desirable phenotype. The plants resulting from adding a gene are often referred to as transgenic plants. If for genetic modification genes of the species or of a crossable plant are used under control of their native promoter, then they are called cisgenic plants. Sometimes genetic modification can produce a plant with the desired trait or traits faster than classical breeding because the majority of the plant's genome is not altered.

To genetically modify a plant, a genetic construct must be designed so that the gene to be added or removed will be expressed by the plant. To do this, a promoter to drive transcription and a termination sequence to stop transcription of the new gene, and the gene or genes of interest must be introduced to the plant. A marker for the selection of transformed plants is also included. In the laboratory, antibiotic resistance is a commonly used marker: Plants that have been successfully transformed will grow on media containing antibiotics; plants that have not been transformed will die. In some instances markers for selection are removed by backcrossing with the parent plant prior to commercial release.

The construct can be inserted in the plant genome by genetic recombination using the bacteria *Agrobacterium tumefaciens* or *A. rhizogenes*, or by direct methods like the gene gun or microinjection. Using plant viruses to insert genetic constructs into plants is also a possibility, but the technique is limited by the host range of the virus. For example, Cauliflower mosaic virus (CaMV) only infects cauliflower and related species. Another limitation of viral vectors is that the virus is not usually passed on the progeny, so every plant has to be inoculated.

The majority of commercially released transgenic plants are currently limited to plants that have introduced resistance to insect pests and herbicides. Insect resistance is achieved through incorporation of a gene from *Bacillus thuringiensis* (Bt) that encodes a protein that is toxic to some insects. For example, the cotton bollworm, a common cotton pest, feeds on Bt cotton it will ingest the toxin and die. Herbicides usually work by binding to certain plant enzymes and inhibiting their action. The enzymes that the herbicide inhibits are known as the herbicides *target site*. Herbicide resistance can be engineered into crops by expressing a version of *target site* protein that is not inhibited by the herbicide. This is the method used to produce glyphosate resistant crop plants.

Genetic modification of plants that can produce pharmaceuticals (and industrial chemicals), sometimes called *pharming*, is a rather radical new area of plant breeding.

Issues and Concerns

Modern plant breeding, whether classical or through genetic engineering, comes with issues of concern, particularly with regard to food crops. The question of whether breeding can have a negative effect on nutritional value is central in this respect. Although relatively little direct research in this area has been done, there are scientific indications that, by favoring certain aspects of a plant's development, other aspects may be retarded. A study published in the *Journal of the American College of Nutrition* in 2004, entitled *Changes in USDA Food Composition Data for 43 Garden Crops, 1950 to 1999*, compared nutritional analysis of vegetables done in 1950 and in 1999, and found substantial decreases in six of 13 nutrients measured, including 6% of protein and 38% of riboflavin. Reductions in calcium, phosphorus, iron and ascorbic acid were also found. The study, conducted at the Biochemical Institute, University of Texas at Austin, concluded in summary: "*We suggest that any real declines are generally most easily explained by changes in cultivated varieties between 1950 and 1999, in which there may be trade-offs between yield and nutrient content.*"

The debate surrounding genetically modified food during the 1990s peaked in 1999 in terms of media coverage and risk perception, and continues today - for example, "*Germany has thrown its weight behind a growing European mutiny over genetically modified crops by banning the planting of a widely grown pest-resistant corn variety.*" The debate encompasses the ecological impact of genetically modified plants, the safety of genetically modified food and concepts used for safety evaluation like substantial equivalence. Such concerns are not new to plant breeding. Most countries have regulatory processes in place to help ensure that new crop varieties entering the marketplace are both safe and meet farmers' needs. Examples include variety registration, seed schemes, regulatory authorizations for GM plants, etc.

Plant breeders' rights is also a major and controversial issue. Today, production of new varieties is dominated by commercial plant breeders, who seek to protect their work and collect royalties through national and international agreements based in intellectual property rights. The range of related issues is complex. In the simplest terms, critics of the increasingly restrictive regulations argue that, through a combination of technical and economic pressures, commercial breeders are reducing biodiversity and significantly constraining individuals (such as farmers) from developing and trading seed on a regional level. Efforts to strengthen breeders' rights, for example, by lengthening periods of variety protection, are ongoing.

When new plant breeds or cultivars are bred, they must be maintained and propagated. Some plants are propagated by asexual means while others are propagated by seeds. Seed propagated cultivars require specific control over seed source and production procedures to maintain the integrity of the plant breeds results. Isolation is necessary to prevent cross contamination with related plants or the mixing of seeds after harvesting. Isolation is normally accomplished by planting distance but in certain crops, plants are enclosed in greenhouses or cages (most commonly used when producing F1 hybrids).

Role of Plant Breeding in Organic Agriculture

Critics of organic agriculture claim it is too low-yielding to be a viable alternative to conventional agriculture. However, part of that poor performance may be the result of growing poorly adapted varieties. It is estimated that over 95% of organic agriculture is based on conventionally adapted varieties, even though the production environments found in organic vs. conventional farming systems are vastly different due to their distinctive management practices. Most notably, organic farmers have fewer inputs available than conventional growers to control their production environments. Breeding varieties specifically adapted to the unique conditions of organic agriculture is critical for this sector to realize its full potential. This requires selection for traits such as:

- Water use efficiency
- Nutrient use efficiency (particularly nitrogen and phosphorus)
- Weed competitiveness
- Tolerance of mechanical weed control
- Pest/disease resistance
- Early maturity (as a mechanism for avoidance of particular stresses)
- Abiotic stress tolerance (i.e. drought, salinity, etc...)

Currently, few breeding programs are directed at organic agriculture and until recently those that did address this sector have generally relied on indirect selection (i.e. selection in conventional environments for traits considered important for organic agriculture). However, because the difference between organic and conventional environments is large, a given genotype may perform very differently in each environment due to an interaction between genes and the environment. If this interaction is severe enough, an important trait required for the organic environment may not be revealed in the conventional environment, which can result in the selection of poorly adapted individuals. To ensure the most adapted varieties are identified, advocates of organic breeding now promote the use of direct selection (i.e. selection in the target environment) for many agronomic traits.

There are many classical and modern breeding techniques that can be utilized for crop improvement in organic agriculture despite the ban on genetically modified organisms. For instance, controlled crosses between individuals allow desirable genetic variation to be recombined and transferred to seed progeny via natural processes. Marker assisted selection can also be employed as a diagnostics tool to facilitate selection of progeny who possess the desired trait(s), greatly speeding up the breeding process. This technique has proven particularly useful for the introgression of resistance genes into new backgrounds, as well as the efficient selection of many resistance genes pyramided into a single individual. Unfortunately, molecular markers are not currently available for many important traits, especially complex ones controlled by many genes.

Addressing Global Food Security Through Plant Breeding

For future agriculture to thrive there are necessary changes which must be made in accordance to arising global issues. These issues are arable land, harsh cropping conditions and food security which involves, being able to provide the world population with food containing sufficient nutrients. These crops need to be able to mature in several environments allowing for worldwide access, this is involves issues such as drought tolerance. These global issues are achievable through the process of plant breeding, as it offers the ability to select specific genes allowing the crop to perform at a level which yields the desired results.

Increased Yield without Expansion

With an increasing population, the production of food needs to increase with it. It is estimated that a 70% increase in food production is needed by 2050 in order to meet the Declaration of the World Summit on Food Security. But with the degradation of agricultural land, simply planting more crops is no longer a viable option. New varieties of plants can in some cases be developed through plant breeding that generate an increase of yield without relying on an increase in land area. An example of this can be seen in Asia, where food production per capita has increased twofold. This has been achieved through not only the use of fertilisers, but through the use of better crops that have been specifically designed for the area.

Breeding for Increased Nutritional Value

Plant breeding can contribute to global food security as it is a cost-effective tool for increasing nutritional value of forage and crops. Improvements in nutritional value for forage crops from the use of analytical chemistry and rumen fermentation technology have been recorded since 1960; this science and technology gave breeders the ability to screen thousands of samples within a small amount of time, meaning breeders could identify a high performing hybrid quicker. The main area genetic increases were made was in vitro dry matter digestibility (IVDMD) resulting in 0.7-2.5% increase, at just 1% increase in IVDMD a single Bos Taurus also known as beef cattle reported 3.2% increase in daily gains. This improvement indicates plant breeding is an essential tool in gearing future agriculture to perform at a more advanced level.

Breeding for Tolerance

Plant breeding of hybrid crops has become extremely popular worldwide in an effort to combat the harsh environment. With long periods of drought and lack of water or nitrogen stress tolerance has become a significant part of agriculture. Plant breeders have focused on identifying crops which will ensure crops perform under these conditions; a way to achieve this is finding strains of the crop that is resistance to drought conditions with low nitrogen. It is evident from this that plant breeding is vital for future agriculture to survive as it enables farmers to produce stress resistant crops hence improving food security.

Participatory Plant Breeding

Participatory plant breeding (PPB) is when farmers are involved in a crop improvement programme with opportunities to make decisions and contribute to the research process at different stages. Participatory approaches to crop improvement can also be applied when plant biotechnologies are being used for crop improvement.

Terms Related to Plant Breeding

- Mentor pollen, inactivated pollen that is compatible with the female plant is mixed with pollen that would normally be incompatible. The mentor pollen has the effect of guiding the foreign pollen to the ovules.
- S1 generation, the product of self-fertilization.

Double-pair Mating

Double-pair mating (DPM) is a mating (crossing) design used in plant breeding. Each individual is mated with two others.

Principles

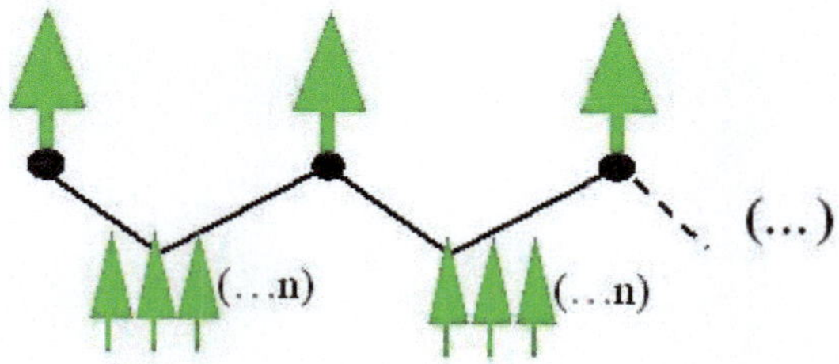

DMP scheme

In figure a connected variant of DPM is shown. DPM is an efficient mating design in balanced breeding programmes, where equal contribution from each breeding population member is desired. With DPM the number of new families created is equal to the number of individuals mated. DPM allows to efficiently utilise positive assortative mating for more efficient use of the breeding population members for deployment to seed orchards. In comparison with single pair mating, DPM has the advantages that the genes from the individual are transmitted to next generation even if one of the crosses fails; that safer estimates of breeding values of the parents get possible (useful for seed orchards, where tested trees are preferred); and that genes from different ancestors have a better chance to combine.

Breeding for Drought Stress Tolerance

Breeding for drought resistance is the process of breeding plants with the goal of reducing the impact of dehydration on plant growth.

Dehydration Stress in Crop Plants

In nature or crop fields, water is often the most limiting factor for plant growth. If plants do not receive adequate rainfall or irrigation, the resulting dehydration stress can reduce growth more than all other environmental stresses combined.

Drought can be defined as the absence of rainfall or irrigation for a period of time sufficient to deplete soil moisture and cause dehydration in plant tissues. Dehydration stress results when water loss from the plant exceeds the ability of the plant's roots to absorb water and when the plant's water content is reduced enough to interfere with normal plant processes.

Dehydration Stress is a Global Phenomenon

About 15 million km2 of the land surface is covered by crop-land, and about 16% of this area is equipped for irrigation (Siebert *et al.* 2005). Thus, in many parts of the world, including the United States, plants may frequently encounter dehydration stress. Rainfall is very seasonal and periodic drought occurs regularly. The effect of drought is more prominent in sandy soils with low water holding capacity. On such soils some plants may experience dehydration stress after only a few days without water.

During the last century, the rate of increase in `blue' water withdrawal (from rivers, lakes, and aquifers) for irrigation and other purposes was higher than the growth rate of the world population (Shiklomanov 1998). Country-wise maps of irrigated areas are available.

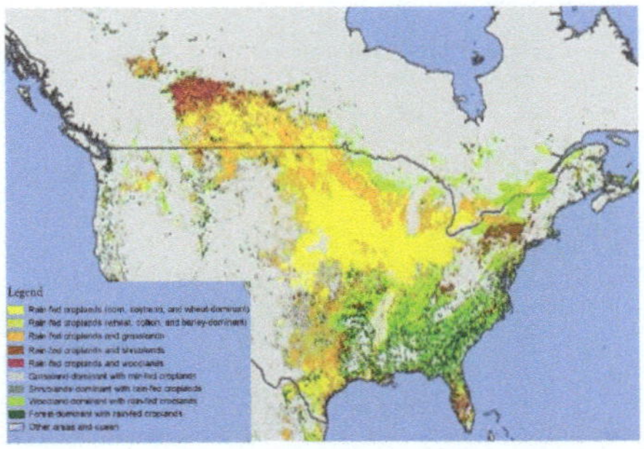

Rain fed areas of USA, for details - Biradar et al.(2009) International J Applied Earth Observation and Geoinformation 11: 114-129

Dehydration Stress and Future Challenges to Crop Production

Soil moisture deficit is a significant challenge to the future of crop production. Severe drought in parts of the U.S., Australia, and Africa in recent years drastically reduced crop yields and disrupted regional economies. Even in average years, however, many agricultural regions, including the U.S. Great Plains, suffer from chronic soil moisture deficits. Cereal crops typically attain only about 25% of their potential yield due to the effects of environmental stress, with dehydration stress the most important cause. Two major trends will likely increase the frequency and severity of soil moisture deficits:

(1) Climate change.

Higher temperatures are likely to increase crop water use due to increased transpiration. A warmer atmosphere will also speed up melting of mountain snow pack, resulting in less water available for irrigation. More extreme weather patterns will increase the frequency of drought in some regions.

(2) Competing uses for limited water supplies.

Increased demand from municipal and industrial users will further reduce the amount of water available for irrigated crops.

Although changes in tillage and irrigation practices can improve production by conserving water, enhancing the genetic tolerance of crops to drought stress is considered an essential strategy for addressing moisture deficits.

Dehydration Stress Affects Plant Physiology

A plant responds to a lack of water by halting growth and reducing photosynthesis and other plant processes in order to reduce water use. As water loss progresses, leaves of some species may appear to change colour — usually to blue-green. Foliage begins to wilt and, if the plant is not irrigated, leaves will fall off and the plant will eventually die. Soil moisture deficit lowers the water potential of a plant's root and, upon extended exposure, abscisic acid is accumulated and eventually stomatal closure occurs. This reduces a plant's leaf relative water content. The time required for dehydration stress to occur depends on the water-holding capacity of the soil, environmental conditions, stage of plant growth, and plant species. Plants growing in sandy soils with low water-holding capacity are more susceptible to dehydration stress than plants growing in clay soils. A limited root system will accelerate the rate at which dehydration stress develops. A plant's root system may be limited by the presence of competing root systems from neighbouring plants, by site conditions such as compacted soils or high water tables, or by container size (if growing in a container). A plant with a large mass of leaves in relation to the root system is prone to drought stress because the leaves may lose water faster than the roots can supply it. Newly planted plants and poorly established plants may be especially susceptible to dehydration stress because of the limited root system or the large mass of stems and leaves in comparison to roots.

Dehydration Stress Interaction with Other Stress Factors

Aside from the moisture content of the soil, environmental conditions of high light intensity, high temperature, low relative humidity and high wind speed will significantly increase plant water loss. The prior environment of a plant also can influence the development of dehydration stress. A plant that has been exposed to dehydration stress (hardened) previously and has recovered may become more drought resistant. Also, a plant that was well-watered prior to being water-limited will usually survive a period of drought better than a continuously dehydration-stressed plant.

Mechanisms of Drought Resistance

The degree of resistance to drought depends upon individual crops. Generally three strategies can help a crop to mitigate the effect of dehydration stress:

The Drought Resistance terms in summary (Levitt, J. (1980); Blum, A. (2011))

Dehydration Avoidance

If the plant shows dehydration avoidance, the environmental factor is excluded from the plant tissues by reducing water loss ("water savers", e.g. by thick leaf epicuticular wax, leaf rolling, leaf posture) or maintaining water uptake ("water spenders", e.g. by growing deeper roots). Dehydration avoidance is desirable in modern agriculture, where drought resistance requires the maintenance of economically viable plant production under dehydration stress. The role of dehydration avoidance is maintaining water supply and sustaining leaf hydration and turgidity with the purpose of maintaining stomatal opening and transpiration as long as possible under water deficit. This is essential for leaf gas exchange, photosynthesis and plant production through carbon assimilation.

Dehydration Tolerance

If the plant shows dehydration tolerance, the environmental factor enters the plant tissues but the tissues survive, by e.g. maintaining turgor and osmotic adjustment.

Dehydration Escape

Dehydration escape involves e.g. early maturing or seed dormancy, where the plant uses previous optimal conditions to develop vigor.

Dehydration Recovery refers to some plant species being able to recuperate after brief drought periods.

A proper timing of life-cycle, resulting in the completion of the most sensitive developmental stages while water is abundant, is considered to be a dehydration escape strategy. Avoiding dehydration stress with a root system capable of extracting water from deep soil layers, or by reducing evapotranspiration without affecting yields, is considered as dehydration avoidance. Mechanisms such as osmotic adjustment (OA) whereby a plant maintains cell turgor pressure under reduced soil water potential are categorised as dehydration tolerance mechanisms. Dehydration avoidance mechanisms can be expressed even in the absence of stress and are then considered constitutive. Dehydration tolerance mechanisms are the result of a response triggered by dehydration stress itself and are therefore considered adaptive. When the stress is terminal and predictable, dehydration escape through the use of shorter duration varieties is often the preferable method of improving yield potential. Dehydration avoidance and tolerance mechanisms are required in situations where the timing of drought is mostly unpredictable.

Drought resistance mechanisms are genetically controlled and genes or QTL responsible for drought resistance have been discovered in several crops which opens avenue for molecular breeding for drought resistance.

Drought Resistance Traits

Resistance to drought is a quantitative trait, with a complex phenotype, often confounded by plant phenology. Breeding for drought resistance is further complicated since several types of abiotic stress, such as high temperatures, high irradiance, and nutrient toxicities or deficiencies can challenge crop plants simultaneously.

(a) Osmotic adjustment When a plant is exposed to water deficit, it may accumulate a variety of osmotically active compounds such as amino acids and sugars, resulting in a lowering of the osmotic potential. Examples of amino acids that may be up-regulated are proline and glycine betaine. This is termed osmotic adjustment and enables the plant to take up water, maintain turgor and survive longer.

(b) Cell membrane stability The ability to survive dehydration is influenced by a cell's ability to survive at reduced water content. This can be considered complementary to OA because both traits will help maintain leaf growth (or prevent leaf death) during water deficit. Crop varieties differ in dehydration tolerance and an important factor for such differences is the capacity of the cell membrane to prevent electrolyte leakage at decreasing water content, or "cell membrane stability (CMS)". The maintenance of membrane function is assumed to mean that cell activity is also maintained. Measurements of CMS have been used in different crops and are known to be correlated with yields under high temperature and possibly under dehydration stress.

(c) Epicuticular wax In sorghum (Sorghum bicolor L. Moench), drought resistance is a trait that is highly correlated with the thickness of the epicuticular wax layer. Experiments have demonstrated that rice varieties with a thick cuticle layer retain their leaf turgor for longer periods of time after the onset of a water-stress.

(d) Partitioning and stem reserve mobilisation.

As photosynthesis becomes inhibited by dehydration, the grain filling process becomes increasingly reliant on stem reserve utilisation. Numerous studies have reported that stem reserve mobilisation capacity is related to yield under dehydration stress in wheat. In rice, a few studies also indicated that this mechanism maintains grain yield under dehydration stress at the grain filling stage. This dehydration tolerance mechanism is stimulated by a decrease in gibberellic acid concentration and an increase in abscisic acid concentration.

(e) Manupulation and Stability of flowering processes.

(f) Seedling drought resistance traits For emergence from deep sowing (to exploit dry upper soil), this is practised to help seedlings reach the receding moisture profile, and to avoid high soil surface temperatures which inhibit germination. Screening at these stage provides practical advantages, specially when managing large amount of germ-plasms.

The Drought Resistant Ideotype

Usually ideotypes are developed to create an ideal plant variety. The following traits constitutes ideotype of wheat by CIMMYT.

1) Large seed size. Helps emergence, early ground cover, and initial biomass.

2) Long coleoptiles. For emergence from deep sowing.

3) Early ground cover.

   Thinner, wider leaves (i.e., with a relatively low specific leaf weight) and a more prostrate growth habit help to increase ground cover, thus conserving soil moisture and potentially increasing radiation use efficiency.

4) High pre-anthesis biomass.

5) Good capacity for stem reserves and remobilisation.

6) High spike photosynthetic capacity.

7) High RLWC/Gs/CTD during grain filling to indicate ability to extract water.

8) Osmotic adjustment.

9) Accumulation of ABA.

   The benefit of ABA accumulation under dehydration has been demonstrated (Innes et al. 1984). It appears to pre-adapt plants to stress by reducing stomatal conductance, rates of cell division, organ size,and increasing development rate. However, high ABA can also result in sterility problems since high ABA levels may abort developing florets.

10) Heat Tolerance. The contribution of heat tolerance to performance under dehydration stress needs to be quantified, but it is relatively easy to screen for (Reynolds et al. 1998).

11) Leaf anatomy: Waxiness, pubescence, rolling, thickness, posture. These traits decrease radiation load to the leaf surface. Benefits include a lower evapotranspiration rate and reduced risk of irreversible photo-inhibition. However, they may also be associated with reduce radiation use efficiency, which would reduce yield under more favourable conditions.

12) High tiller survival.

    Comparison of old and new varieties have shown that under dehydration older varieties over-produce tillers many of which fail to set grain while modern drought resistant lines produce fewer tillers most of which survive.

13) Stay-green.

The trait may indicate the presence of drought resistance mechanisms, but probably does not contribute to yield per se if there is no water left in the soil profile by the end of the cycle to support leaf gas exchange. It may be detrimental if it indicates lack of ability to remobilise stem reserves. However, research in sorghum has indicated that Stay-green is associated with higher leaf chlorophyll content at all stages of development and both were associated with improved yield and transpiration efficiency under dehydration.

Combination Phenomics: Overall Health of Crops

The concept of combination phenomics comes from the idea that two or more plant stresses have common physiological effects or common traits - which are an indicator of overall plant health. As both biotic and abiotic stresses can result in similar physiological consequence, drought resistant plants can be separated from sensitive plants. Some imaging or infrared measuring techniques can help to speed the process for breeding process. For example spot blotch intensity and canopy temperature depression can be monitored with canopy temperature depression.

Molecular Breeding for Drought Resistance

Recent research breakthroughs in biotechnology have revived interest in targeted drought resistance breeding and use of new genomics tools to enhance crop water productivity. Marker-assisted breeding is revolutionising the improvement of temperate field crops and will have similar impacts on breeding of tropical crops. Other molecular breeding tool include development of genetically modified crops that can tolerate plant stress. As a complement to the recent rapid progress in genomics, a better understanding of physiological mechanisms of dehydration response will also contribute to the progress of genetic enhancement of crop drought resistance. It is now well accepted that the complexity of the dehydration syndrome can only be tackled with a holistic approach that integrates physiological dissection of crop dehydration avoidance and - tolerance traits using molecular genetic tools such as MAS, micro-arrays and transgenic crops, with agronomic practices that lead to better conservation and utilisation of soil moisture, and better matching of crop genotypes with the environment.

Breeding for Heat Stress Tolerance

Plant breeding is process of development of new cultivars. Plant breeding involves development of varieties for different environmental conditions – some of them are not favorable. Among them, heat stress is one of such factor that reduces the production and quality significantly. So breeding against heat is a very important criterion for breeding for current as well as future environments produced by global climate change (e.g. global warming).

Breeding for Heat Stress Tolerance in Plants

Heat stress due to increased temperature is a very important problem globally. Occasional or prolonged high temperatures cause different morpho-anatomical, physiological and biochemical changes in plants. The ultimate effect is on plant growth as well as development and reduced yield and quality. Breeding for heat stress tolerance can be mitigated by breeding plant varieties that have improved levels of thermo-tolerance using different conventional or advanced genetic tools.

What is Heat Stress Tolerance

Heat stress is defined as increased temperature level sufficient to cause irreversible damage to plant growth and development. Generally a temperature rise, above usually 10 to 15 °C above ambient, can be considered heat shock or heat stress. Heat tolerance is broadly defined as the ability of the plant tolerate heat – means that grow and produce economic yield under high temperatures.

Significance: Current and Future - global Warming

Heat stress is a serious threat to crop production globally (Hall, 2001, 1992). Global warming is particularly consequence of increased level of green house gases such as CO_2, methane, chlorofluorocarbons and nitrous oxides. The Intergovernmental Panel on Climatic Change (IPCC) has predicted a rise of 0.3 °C per decade (Jones et al., 1999) reaching to approximately 1 and 3 °C above the present value by 2025 and 2100 AD, respectively.

Physiological Consequence of Heat Stress

At very high temperatures cause severe cellular injury and cell death may occur within short time, thus leading to a catastrophic collapse of cellular organization (Schoffl et al., 1999). However, under moderately high temperatures, the injury can only occur after longer exposure to such a temperature however the plant efficiency can be severely affected. High temperature directly affect injuries such as protein denaturation and aggregation, and increased fluidity of membrane lipids. Other indirect or slower heat injuries involve inactivation of enzymes in chloroplast and mitochondria, protein degradation, inhibition of protein synthesis, and loss of membrane integrity. Heat stress associated injuries ultimately lead to starvation, inhibition of growth, reduced ion flux, production of toxic compounds and production of reactive oxygen species (ROS). Immediately after exposure to high temperature stress-related proteins are expressed as stress defense strategy of the cell. Expression of heat shock proteins (HSPs), protein with 10 to 200 kDa, is supposed to be involved in signal transduction during heat stress. In many species it has been demonstrated that HSPs results in improved physiological phenomena such as photosynthesis, assimilate partitioning, water and nutrient use efficiency, and membrane stability.

Studies have found tremendous variation within and between species, thus this will help to breed heat tolerance for future environment. Some of attempts to develop heat-tolerant genotypes are successful. (Ehlers and Hall, 1998; Camejo et al., 2005).

Traits Associated with Heat Stress Tolerance

Different physiological mechanisms may contribute to heat tolerance in the field—for example, heat tolerant metabolism as indicated by higher photosynthetic rates, stay-green, and membrane thermo-stability, or heat avoidance as indicated by canopy temperature depression. Several physiological and morphological traits have been evaluated for heat tolerance - Canopy temperature, leaf chlorophyll, stay green, leaf conductance, spike number, biomass, and flowering date.

(A) Canopy temperature depression (CTD)

CTD has shown clear association with yield in warm environments shows it association with heat stress tolerance. CTD shows high genetic correlation with yield and high values of proportion of direct response to selection (Reynolds et al., 1998) indicating that the trait is heritable and therefore amenable to early generation selection. Since an integrated CTD value can be measured almost instantaneously on scores of plants in a small breeding plot (thus reducing error normally associated with traits measured on individual plants), work has been conducted to evaluate its potential as an indirect selection criterion for genetic gains in yield. CTD is affected by many physiological factors, which makes it a powerful.

(B) Stomatal conductance

Canopy temperature depression is highly suitable for selecting physiologically superior lines in warm, low relative humidity environments where high evaporative demand leads to leaf cooling of up to 10 °C below ambient temperatures. This permits differences among genotypes to be detected relatively easily using infrared thermometry. However, such differences cannot be detected in high relative humidity environments because the effect of evaporative cooling of leaves is negligible. Nonetheless, leaves maintain their stomata open to permit the uptake of CO_2, and differences in the rate of CO_2 fixation may lead to differences in leaf conductance that can be measured using a porometer. Porometry can be used to screen individual plants. The heritability of stomatal conductance is reasonably high, with reported values typically in the range of 0.5 to 0.8. Plants can be assessed for leaf conductance using a viscous flow porometer that is available on the market (Thermoline and CSIRO, Australia). This instrument can give a relative measure of stomatal conductance in a few seconds, making it possible to identify physiologically superior genotypes from within bulks.

(C) Membrane thermostability

Although resistance to high temperatures involves several complex tolerance and avoidance mechanisms, the membrane is thought to be a site of primary physiological injury by heat, and measurement of solute leakage from tissue can be used to estimate damage to membranes. Since membrane thermostability is reasonably heritable (Fokar et al., 1998) and shows high genetic correlation with yield.

(D) Chlorophyll fluorescence

Chlorophyll fluorescence, an indication of the fate of excitation energy in the photosynthetic apparatus, has been used indicator for heat stress tolerance.

(E) Chlorophyll content and stay green

Chlorophyll content and stay green traits have been found to be associated with heat stress tolerance.,. Xu et al. (2000) identified three QTLs for chlorophyll content (Chl1, Chl2, and Chl3) (coincided with three stay-green QTL regions (Stg1, Stg2, and Stg3)) were identified in Sorghum. The Stg1 and Stg2 regions also contain the genes for key photosynthetic enzymes, heat shock proteins, and an abscisic acid (ABA) responsive gene.

(F) Photosynthesis

Decliend photosynthesis is suggested as measure of heat stress sensitivity in plants.

(G) Stem reserve remobilization.

Combination Breeding and Physiological Breeding

The physiological-trait-based breeding approach has merit over breeding for yield *per se* because it increases the probability of crosses resulting in additive gene action. The concept of combination phenomics comes from the idea that two or more stress have common physiological effect or common traits - which is an indicator of overall plant health. Similar analogy in human medical terms is high blood pressure or high body temperature or high white blood cells in body is an indicator of health problems and thus we can select healthy people from unhealthy using such a measure. As both abiotic and abiotic stresses can result in similar physiological consequence, tolerant plant can be separated from sensitive plants. Some imaging or infrared measuring techniques can help to speed the process for breeding process. For example, spot blotch intensity and canopy temperature depression can be monitored with canopy temperature depression.

Breeding of Strawberries

The breeding of strawberries is a long story that starts with the selection and cultivation of European strawberry species in western Europe in the 15th century while a similar discovery and cultivation occurred in Chile. The most commonly consumed strawberry species in modern times is the garden strawberry, a species derived from hybridization of two other species, with the scientific name *Fragaria × ananassa*, but there are many species of strawberries, several others of which are cultivated to some extent. The strawberry species fall into several different genetic types, based on their number of chromosomes. Strawberry growers have employed many breeding techniques, starting with traditional plant breeding and then moving on to molecular breeding and genetic engineering in the 20th century.

Watercolor picture of a strawberry plant by Deborah Griscom Passmore in 1890

The History of the Breeding of Strawberries

European Breeding Before American Contact

In Europe, there were three main strawberry species. They are *Fragaria vesca, F. viridis* and *F. moschata*. By the time American plants were introduced, the most commonly grown of these was *F. vesca*.

Fragaria Vesca

A species that has both red and blue varieties. It is also known as the "woodland strawberry". Of particular interest is *F. sylvestris* var. *semperflorens*, a variety of *F vesca*. *F. sylvestris* var. *semperflorens* is unusual, because it is ever-bearing, meaning that it constantly flowers and bears fruit until frosts are too much for it in the autumn.

Fragaria Viridis

This species is also known as a "green strawberry". Its origins are in the Alps. A unique characteristic of this plant is that it is twice bearing, meaning that it will flower and bear fruit twice in a year.

Fragaria Moschata

Also known as "musky flavoured strawberry" or "Hautboy", in England. This species was known for having large fruit with a slightly musky scent.

France

The origins of modern large-fruited strawberries can be found in France. In 1714 *Fragaria chiloensis*, a plant that produces large fruit that is particularly good for eating, was taken from South America to France by a French spy. After its arrival in France, this variety was bred with *Fragaria virginiana*, a hearty plant from North America. The product of this cross is the species *Fragaria × ananassa*.

Antoine Nicolas Duchesne

Antoine Nicolas Duchesne is a very important character in the development of strawberries in both France and the rest of the world. He discovered that strawberries can be either bisexual or unisexual. He also conducted experiments, crossing *F. moschata* and *F. chiloensis*. The resulting large fruit put Duchesne in King Louis XV's favor and allowed him to continue to study and create his categorization of the ten "races" of strawberry.

England

While France did much to establish modern large fruit strawberries, they were not alone in their development of the plant. The most commonly used strawberry was *Fragaria virginiana* or the "scarlet strawberry". It was commonly used because of the English holdings in North America, *F. virginiana*'s genetic home. The English work with breeding was centered on breeding *F. virginiana* into new varieties and crossing it with *F. chiloensis*. The reason for this was that *F. chiloensis* has a large berry size and a pleasant flavor but poor resistance to the climate of England. The English breeding of early *F. virginiana* x *F. chiloensis* crosses can be examined by looking at two of the most successful breeders of England: Andrew Knight and Michael Keens.

Andrew Knight

Prior to being one of the founding member of the Royal Horticulture Society in 1804, Thomas Andrew Knight generally refused to read any kind of paper concerning his research interest or publish any of his own findings. However, he eventually became part of the academic community and England benefited greatly from it. His work was mostly with different kinds of *F. virginiana* x *F. chiloensis* plants. While he developed many successful varieties in his 1817 breeding experiment, he was mistaken in his belief that all inter-fertile large fruited strawberries were the same species.

Michael Keens

Michael Keens was far less methodical than Knight. He did develop an extremely popular variety that was praised for its large size and excellent flavor up until the 20th century.

Modern Breeding

In the modern age, strawberry breeding is a delicate science and art. Its aim is to produce plant varieties that will be able to supply the world's demand for fruit by overcoming adverse conditions and disease.

United States

In 1920 a large shift in the breeding of strawberries occurred. Breeding stopped being a largely private personal endeavor and became a government matter when the United States department of Agriculture started to fund strawberry breeding. The early objectives of the breeding stations were to develop new varieties to better satisfy the American demand for better dessert, canning and freezing varieties. In the late 1930s and 40s disease resistance became an objective of breeding, particularly to red stele root disease. The 1937 federal work objectives included goals of improving resistance to disease and improving tolerance of long and short days as well as high and low temperatures.

Great Britain

By the 1950s many American varieties were being used in Great Britain. In the mid 50s, varieties were released that were resistant to red stele root disease.

France

Unlike Great Britain and the United States, France allowed strawberry breeding to remain a largely private study for the early 20th century. This allowed individual breeders to follow their own curiosity and work with plants that were atypical for the time. A notable example is Charles Simmen's work with ever-bearing strawberries.

Private Sector Funded Breeding

As part of research and development, many agricultural businesses have seen fit to invest in the development of their own varieties.

Genetics of Strawberries

Strawberries have many different chromosome numbers. While these are four of the most common numbers of chromosome pairs some strawberries can have as many as 16.

Diploid

- *Fragaria daltoniana* — native to Asia
- *Fragaria nilgerrensis* — native to South Asia
- *Fragaria nubicola* — native to South Asia
- *Fragaria vesca* — found throughout Europe, North America, and Northern Asia as well as North Africa, the mountains of South America, and the northern polar regions

- *Fragaria viridis* — native to Central Europe

Tetraploid

- *Fragaria moupinensis* — native to East Central Asia
- *Fragaria orientalis* — native to Northeast Asia

Hexaploid

- *Fragaria moschata* — native to Central Europe

Octaploid

- *Fragaria chiloensis* — native to South Chile,mountains of Hawaii
- *Fragaria ovalis* — native to Western North America
- *Fragaria virginiana* — native to Eastern North America

The Breeding Techniques

Traditional Breeding

Traditional breeding refers to the process of allowing certain chosen plants to sexually reproduce with other plants. Plants are chosen based on favorable characteristics. Simply put, traditional breeding takes plants with favorable characteristics and breeds them. Then the offspring are raised and then judgment is made about which ones have the best traits and the process proceeds to the next generation. This method has been the way that humans have traditionally modified organisms. Not until the 20th century were humans able to influence the genotypes of organisms in any other way.

An Example Variety Developed by Andrew Knight

The "Downton" was a successful variety developed by Andrew Knight as a result of his 1817 breeding experiment. The mother of this variety was a plant grown from seeds direct from America (probably F. Virginiana) and its father was the variety "Old Black", which is of uncertain origin. This variety was created by pollination, not direct manipulation of the plant's genes.

Molecular Breeding

Molecular breeding is the application of molecular biology tools in a breeding program.

Genetic Engineering

Genetic engineering is typically defined as the introduction of non-native genes into an organism.

An Example of Cold Resistance

An excellent example of transgenic modification is in the case of cold resistant strawberries. In one particular variety genes from the arctic flounder, a fish that lives in very cold water, were used to give plants resistance to cold. This modification works because of the genetics of the arctic flounder. It lives in water where other fish would freeze to death but, with a special gene that allows it to produce a sort of anti-freeze, it can survive. This gene is put into a bacteria that is sprayed on the strawberry during the freezing temperatures, allowing it to also be resistant to cold. The strawberry is then cleaned, removing the bacteria.

Doubled Haploidy

A doubled haploid (DH) is a genotype formed when haploid cells undergo chromosome doubling. Artificial production of doubled haploids is important in plant breeding.

Haploid cells are produced from pollen or egg cells or from other cells of the gametophyte, then by induced or spontaneous chromosome doubling, a doubled haploid cell is produced, which can be grown into a doubled haploid plant. If the original plant was diploid, the haploid cells are monoploid, and the term doubled monoploid may be used for the doubled haploids. Haploid organisms derived from tetraploids or hexaploids are sometimes called dihaploids (and the doubled dihaploids are, respectively, tetraploid or hexaploid).

Conventional inbreeding procedures take six generations to achieve approximately complete homozygosity, whereas doubled haploidy achieves it in one generation. Dihaploid plants derived from tetraploid crop plants may be important for breeding programs that involve diploid wild relatives of the crops.

History

The first report of the haploid plant was published by Blakeslee *et al.* (1922) in *Datura stramonium*. Subsequently, haploids were reported in many other species. Guha and Maheshwari (1964) developed an anther culture technique for the production of haploids in the laboratory. Haploid production by wide crossing was reported in barley (Kasha and Kao, 1970) and tobacco (Burk *et al.*, 1979). Tobacco, rapeseed, and barley are the most responsive species for doubled haploid production. Doubled haploid methodologies have now been applied to over 250 species.

Production of Doubled Haploids

Doubled haploids can be produced *in vivo* or *in vitro*. Haploid embryos are produced *in vivo* by parthenogenesis, pseudogamy, or chromosome elimination after wide crossing. The haploid embryo is rescued, cultured, and chromosome-doubling produces doubled haploids. The *in vitro* methods include gynogenesis (ovary and flower culture) and androgenesis (anther and microspore culture). Androgenesis is the preferred method. Another method of producing the haploids is wide crossing. In barley, haploids can be produced by wide crossing with the related species *Hordeum bulbosum*; fertilization is affected, but during the early stages of seed development the *H. bulbosum* chromosomes are eliminated leaving a haploid embryo. In tobacco (*Nicotiana tabacum*), wide crossing with *Nicotiana africana* is widely used. When *N. africana* is used to pollinate *N.*

tabacum, 0.25 to 1.42 percent of the progeny survive and can readily be identified as either F1 hybrids or maternal haploids. Although these percentages appear small, the vast yield of tiny seeds and the early death of most seedlings provide significant numbers of viable hybrids and haploids in relatively small soil containers. This method of interspecific pollination serves as a practical way of producing seed-derived haploids of *N. tabacum*, either as an alternative method or complementary method to anther culture.

Genetics of DH Population

In DH method only two types of genotypes occur for a pair of alleles, A and a, with the frequency of ½ AA and ½ aa, while in diploid method three genotypes occur with the frequency of ¼ AA, ½ Aa, ¼ aa. Thus, if AA is desirable genotype, the probability of obtaining this genotype is higher in haploid method than in diploid method. If n loci are segregating, the probability of getting the desirable genotype is (1/2)n by the haploid method and (1/4)n by the diploid method. Hence the efficiency of the haploid method is high when the number of genes concerned is large.

Studies were conducted comparing DH method and other conventional breeding methods and it was concluded that adoption of doubled haploidy does not lead to any bias of genotypes in populations, and random DHs were even found to be compatible to selected line produced by conventional pedigree method.

Applications of DHs Plant Breeding

Mapping Quantitative Trait Loci

Most of the economic traits are controlled by genes with small but cumulative effects. Although the potential of DH populations in quantitative genetics has been understood for some time, it was the advent of molecular marker maps that provided the impetus for their use in identifying loci controlling quantitative traits. As the quantitative trait loci (QTL) effects are small and highly influenced by environmental factors, accurate phenotyping with replicated trials is needed. This is possible with doubled haploidy organisms because of their true breeding nature and because they can conveniently be produced in large numbers. Using DH populations, 130 quantitative traits have been mapped in nine crop species. In total, 56 DH populations were used for QTL detection.

Backcross Breeding

In backcross conversion, genes are introgressed from a donor cultivar or related species into a recipient elite line through repeated backcrossing. A problem in this procedure is being able to identify the lines carrying the trait of interest at each generation. The problem is particularly acute if the trait of interest is recessive, as it will be present only in a heterozygous condition after each backcross. The development of molecular markers provides an easier method of selection based on the genotype (marker) rather than the phenotype. Combined with doubled haploidy it becomes more effective. In marker assisted backcross conversion, a recipient parent is crossed with a donor line and the hybrid (F1) backcrossed to the recipient. The resulting generation (BC1) is backcrossed and the process repeated until the desired genotypes are produced. The combination of doubled haploidy and molecular marker provides the short cut. In the back cross generation one itself a genotype with the character of interest can be selected and converted into homozygous doubled haploid genotype. Chen *et al.* (1994) used marker assisted backcross conversion with doubled haploidy of BC1 individuals to select stripe rust resistant lines in barley.

Bulked Segregant Analysis (BSA)

In bulked segregant analysis, a population is screened for a trait of interest and the genotypes at the two extreme ends form two bulks. Then the two bulks are tested for the presence or absence of molecular markers. Since the bulks are supposed to contrast in the alleles that contribute positive and negative effects, any marker polymorphism between the two bulks indicates the linkage between the marker and trait of interest. BSA is dependent on accurate phenotyping and the DH population has particular advantage in that they are true breeding and can be tested repeatedly. DH populations are commonly used in bulked segregant analysis, which is a popular method in marker assisted breeding. This method has been applied mostly to rapeseed and barley.

Genetic Maps

Genetic maps are very important to understand the structure and organization of genomes from which evolution patterns and syntenic relationships between species can be deduced. Genetic maps also provide a framework for the mapping of genes of interest and estimating the magnitude of their effects and aid our understanding of genotype/phenotype associations. DH populations have become standard resources in genetic mapping for species in which DHs are readily available. Doubled haploid populations are ideal for genetic mapping. It is possible to produce a genetic map within two years of the initial cross regardless of the species. Map construction is relatively easy using a DH population derived from a hybrid of two homozygous parents as the expected segregation ratio is simple, *i.e.* 1:1. DH populations have now been used to produce genetic maps of barley, rapeseed, rice, wheat, and pepper. DH populations played a major role in facilitating the generation of the molecular marker maps in eight crop species.

Genetic Studies

Genetic ratios and mutation rates can be read directly from haploid populations. A small doubled haploid (DH) population was used to demonstrate that a dwarfing gene in barley is located chromosome 5H. In another study the segregation of a range of markers has been analyzed in barley.

Genomics

Although QTL analysis has generated a vast amount of information on gene locations and the magnitude of effects on many traits, the identification of the genes involved has remained elusive. This is due to poor resolution of QTL analysis. The solution for this problem would be production of recombinant chromosome substitution line, or stepped aligned recombinant inbred lines. Here, backcrossing is carried out until a desired level of recombination has occurred and genetic markers are used to detect desired recombinant chromosome substitution lines in the target region, which can be fixed by doubled haploidy. In rice, molecular markers have been found to be linked with major genes and QTLs for resistance to rice blast, bacterial blight, and sheath blight in a map produced from DH population.

Elite Crossing

Traditional breeding methods are slow and take 10–15 years for cultivar development. Another disadvantage is inefficiency of selection in early generations because of heterozygosity. These two disadvantages can be over come by DHs, and more elite crosses can be evaluated and selected within less time.

Cultivar Development

Uniformity is a general requirement of cultivated line in most species, which can be easily obtained through DH production. There are various ways in which DHs can be used in cultivar production. The DH lines themselves can be released as cultivars, they may be used as parents in hybrid cultivar production or more indirectly in the creation of breeders lines and in germplasm conservation. Barley has over 100 direct DH cultivars. According to published information there are currently around 300 DH derived cultivars in 12 species worldwide.

The relevance of DHs to plant breeding has increased markedly in recent years owing to the development of protocols for 25 species. Doubled haploidy already plays an important role in hybrid cultivar production of vegetables, and the potential for ornamental production is being vigorously examined. DHs are also being developed in the medicinal herb *Valeriana officinalis* to select lines with high pharmacological activity. Another interesting development is that fertile homozygous DH lines can be produced in species that have self-incompatibility systems.

Advantages of DHs

The ability to produce homozygous lines after a single round recombination saves a lot of time for the plant breeders. Studies conclude that random DH's are comparable to the selected lines in pedigree inbreeding. The other advantages include development of large number of homozygous lines, efficient genetic analysis and development of markers for useful traits in much less time. More specific benefits include the possibility of seed propagation as an alternative to vegetative multiplication in ornamentals, and in species such as trees in which long life cycles and inbreeding depression preclude traditional breeding methods, doubled haploidy provides new alternatives.

Disadvantages of DHs

The main disadvantage with the DH population is that selection cannot be imposed on the population. But in conventional breeding selection can be practised for several generations: thereby desirable characters can be improved in the population.

In haploids produced from anther culture, it is observed that some plants are aneuploids and some are mixed haploid-diploid types. Another disadvantage associated with the double haploidy is the cost involved in establishing tissue culture and growth facilities. The over-usage of doubled haploidy may reduce genetic variation in breeding germplasm. Hence one has to take several factors into consideration before deploying doubled haploidy in breeding programmes.

Technological advances have now provided DH protocols for most plant genera. The number of species amenable to doubled haploidy has reached a staggering 250 in just a few decades. Response efficiency has also improved with gradual removal of species from recalcitrant category. Hence it will provide greater efficiency of plant breeding.

Plant Propagation

Gentian seedlings in a plant nursery

Plant propagation is the process of creating new plants from a variety of sources: seeds, cuttings, bulbs and other plant parts. Plant propagation can also refer to the artificial or natural dispersal of plants.

Sexual Propagation (Seed)

One way to propagate an avocado seed

Seeds and spores can be used for reproduction (through e.g. sowing). Seeds are typically produced from sexual reproduction within a species, because genetic recombination has occurred. A plant grown from seeds may have different characteristics from its parents. Some species produce seeds that require special conditions to germinate, such as cold treatment. The seeds of many Australian plants and plants from southern Africa and the American west require smoke or fire to germinate. Some plant species, including many trees do not produce seeds until they reach maturity, which may take many years. Seeds can be difficult to acquire and some plants do not produce seed at all.

Some plants (like certain Plants modified using genetic use restriction technology may produce seed, but not fertile seed. In certain cases, this is done to prevent the accidental spreading of these plants, for example by birds and other animals.

Asexual Propagation

Plants have a number of mechanisms for asexual or vegetative reproduction. Some of these have been taken advantage of by horticulturists and gardeners to multiply or clone plants rapidly. Humans may utilize these processes as propagation methods, such as tissue culture and grafting. Plants are produced using material from a single parent and as such there is no exchange of genetic material, therefore vegetative propagation methods almost always produce plants that are identical to the parent. Vegetative reproduction uses plants parts such as roots, stems and leaves. In some plants seeds can be produced without fertilization and the seeds contain only the genetic material of the parent plant. Therefore, propagation via asexual seeds or apomixis is asexual reproduction but not vegetative propagation.

Softwood stemcuttings rooting in a controlled environment

Techniques for vegetative propagation include:

- Air or ground layering
- Division
- Grafting and bud grafting, widely used in fruit tree propagation
- Micropropagation
- Stolons or runners
- Storage organs such as bulbs, corms, tubers and rhizomes
- Striking or cuttings
- Twin-scaling
- Offsets

Heated Propagator

A heated propagator is a horticultural device to maintain a warm and damp environment for seeds and cuttings to grow in.

This can be in the form of a clear enclosed bin sitting over a hotpad, or even a portable heater pointed at the bin. The key is to keep the moisture in the clear bin, while keeping lighting over the top of it, usually.

Seed Propagation Mat

An electric seed-propagation mat is a heated rubber mat covered by a metal cage which is used in gardening. The mats are made so that planters containing seedlings can be placed on top of the metal cage without the risk of starting a fire. In extreme cold, gardeners place a loose plastic cover over the planters/mats which creates a sort of miniature greenhouse. The constant and predictable heat allows people to garden in the winter months when the weather is generally too cold for seedlings to survive naturally. When combined with a lighting system, many plants can be grown indoors using these mats.

Cultivar

Osteospermum 'Pink Whirls'
A cultivar selected for its intriguing and colourful flowers

The term *cultivar* most commonly refers to an assemblage of plants selected for desirable characteristics that are maintained during propagation. More generally, *cultivar* refers to the most basic classification category of cultivated plants governed by the ICNCP. Most cultivars have arisen in cultivation, but a few are special selections from the wild.

Popular ornamental garden plants like roses, camellias, daffodils, rhododendrons, and azaleas are cultivars produced by careful breeding and selection for flower colour and form. Similarly, the world's agricultural food crops are almost exclusively cultivars that have been selected for characteristics such as improved yield, flavour, and resistance to disease and very few wild plants are now used as food sources. Trees used in forestry are also special selections grown for their enhanced quality and yield of timber.

Cultivars form a major part of Liberty Hyde Bailey's broader grouping, the cultigen, defined as a plant whose origin or selection is primarily due to intentional human activity. *Cultivar* was coined by Bailey and it is generally regarded as a portmanteau of "cultivated" and "variety", but could also be derived from "cultigen" and "variety". A cultivar is not the same as a botanical variety, a taxonomic rank below subspecies, and there are differences in the rules for creating and using the names of botanical varieties and cultivars. In recent times, the naming of cultivars has been complicated by the use of statutory plant patents and plant breeders' rights names.

The International Union for the Protection of New Varieties of Plants (UPOV – French: *Union internationale pour la protection des obtentions végétales*) offers legal protection of plant cultivars to people or organisations who introduce new cultivars to commerce. UPOV requires that a cultivar be *distinct, uniform* and *stable*. To be distinct, it must have characteristics that easily distinguish it from any other known cultivar. To be uniform and stable, the cultivar must retain these characteristics under repeated propagation.

The naming of cultivars is an important aspect of cultivated plant taxonomy, and the correct naming of a cultivar is prescribed by the Rules and Recommendations of the *International Code of Nomenclature for Cultivated Plants* (the *ICNCP*, commonly known as the *Cultivated Plant Code*). A cultivar is given a cultivar name, which consists of the scientific Latin botanical name followed by a cultivar epithet. The cultivar epithet is usually in a vernacular language. For example, the full cultivar name of the King Edward potato is *Solanum tuberosum* 'King Edward'. The 'King Edward' part of the name is the cultivar epithet, which, according to the Rules of the *Cultivated Plant Code*, is bounded by single quotation marks.

Origin of Term

Liberty Hyde Bailey (1858–1954) coined the word *cultigen* in 1918 and *cultivar* in 1923.

The origin of the term "cultivar" arises from the need to distinguish between wild plants and those with characteristics that have arisen in cultivation (what we now call cultigens). This distinction dates back to the Greek philosopher Theophrastus (370–285 BCE), the "Father of Botany", who was keenly aware of this difference. Botanical historian Alan Morton notes that Theophrastus in his *Enquiry into Plants* "*had an inkling of the limits of culturally induced (phenotypic) changes and of the importance of genetic constitution*" (*Historia Plantarum* III, 2,2 and *Causa Plantarum* I, 9,3).

The International Code of Nomenclature for algae, fungi, and plants uses as its starting point for modern botanical nomenclature those Latin names that appeared in Linnaeus' publications *Species Plantarum* (10th ed.) and *Genera Plantarum* (5th ed.). In *Species Plantarum*, Linnaeus (1707–1778) listed all the plants known to him, either directly or from his extensive reading. He recognised the rank of varietas (in English this is the botanical "variety", a rank below that of species and subspecies) and he indicated these varieties by using letters of the Greek alphabet such as α, β, λ in front of the variety name, rather than using the abbreviation var., which is the current convention. Most of the varieties listed by Linnaeus were of "garden" origin rather than being wild plants.

Over time there was an increasing need to distinguish between plants growing in the wild, and those with variations that had been produced in cultivation. In the nineteenth century many "garden-derived" plants were given horticultural names, sometimes in Latin and sometimes in a local language. From about the 1900s, plants produced in cultivation in Europe were recognised in the Scandinavian, Germanic, and Slavic literature through the words *stamm* or *sorte* but these words could not be used internationally since, by international agreement, any new terms had to be based in Latin. In the twentieth century an improved international terminology was proposed for the classification and nomenclature of cultivated plants.

The word *cultivar* was coined in 1923 by Liberty Hyde Bailey of Cornell University, New York State, when he wrote:

The cultigen is a species, or its equivalent, that has appeared under domestication – the plant is cultigenous. I now propose another name, cultivar, for a botanical variety, or for a race subordinate to species, that has originated under cultivation; it is not necessarily, however, referable to a recognized botanical species. It is essentially the equivalent of the botanical variety except in respect to its origin.

In this paper Bailey used only the rank of species for the cultigen but it was clear to him that many domesticated plants were more like botanical varieties than species, and that appears to have motivated the suggestion of the new classification category *cultivar*, which is generally assumed to be a contraction of the words *cultivated* and *variety*. However, Bailey was never explicit about the etymology of the word, and it has been suggested that it is a contraction of the words *cultigen* and *variety*, which seems more appropriate.

The new word *cultivar* was promoted as "euphonious" and "free from ambiguity". It serves a purpose. Its use was subsequently recommended by the first *Cultivated Plant Code*, which was published in 1953, and by 1960 it had achieved wide international acceptance.

Cultigens

The terms *cultigen* and *cultivar* may be confused with each other. A *cultigen* is any plant deliberately altered or selected for cultivation by humans; the *Cultivated Plant Code* states that cultigens are "maintained as recognisable entities solely by continued propagation". A cultigen or a component of a cultigen, which is recognisable and stable, can be accepted as a *cultivar*.

Bread wheat, *Triticum aestivum*, is considered a cultigen, and is a distinct species from other wheats according to the biological species concept. Many separate cultivars have been created within the cultigen. Many other cultigens are not considered to be distinct species, and can be named in other ways.

Cultigens can have names at any of various ranks, including cultivar names, names in the classification categories of grex and group, binomial Latin species names, variety names, forma names, and they may be plants that have been altered by humans (including genetically modified plants) but which have not been given formal names.

Formal Definition

The *Cultivated Plant Code* notes that the word cultivar is used in two different senses: first, as a "classification category" the cultivar is defined in Article 2 of the *International Code of Nomenclature for Cultivated Plants* (2009, 8th edition) as follows: *The basic category of cultivated plants whose nomenclature is governed by this Code is the cultivar*. There are two other classification categories for cultigens, the grex and the group. The *Code* then defines a *cultivar* as a "taxonomic unit within the classification category of cultivar". This is the sense of *cultivar* that is most generally understood and which is used as a general definition.

A cultivar is an assemblage of plants that (a) has been selected for a particular character or combination of characters, (b) is distinct, uniform and stable in those characters, and (c) when propagated by appropriate means, retains those characters.

Different Kinds

A cultivar of the orchid genus *Oncidium*

Which plants are chosen to be named as cultivars is simply a matter of convenience as the category was created to serve the practical needs of horticulture, agriculture, and forestry.

Members of a particular cultivar are not necessarily genetically identical. The *Cultivated Plant Code* emphasizes that different cultivated plants may be accepted as different cultivars, even if they have the same genome, while cultivated plants with different genomes may be regarded as the same cultivar. The production of cultivars generally entails considerable human involvement although in a few cases it may be as little as simply selecting variation from plants growing in the wild (whether by collecting growing tissue to propagate from or by gathering seed).

Cultivars generally occur as ornamentals and food crops: *Malus* 'Granny Smith' and *Malus* 'Red Delicious' are cultivars of apples propagated by cuttings or grafting, *Lactuca* 'Red Sails' and *Lactuca* 'Great Lakes' are lettuce cultivars propagated by seeds. Named cultivars of *Hosta* and *Hemerocallis* plants are cultivars produced by micropropagation or division.

Clones

Leucospermum 'Scarlet Ribbon'

A cross performed in Tasmania between *L. glabrum* and *L. tottum*

Cultivars that are produced asexually are genetically identical and known as clones; this includes plants propagated by division, layering, cuttings, grafts, and budding. The propagating material may be taken from a particular part of the plant, such as a lateral branch, or from a particular phase of the life cycle, such as a juvenile leaf, or from aberrant growth as occurs with witch's broom. Plants whose distinctive characters are derived from the presence of an intracellular organism may also form a cultivar provided the characters are reproduced reliably from generation to generation. Plants of the same chimera (which have mutant tissues close to normal tissue) or graft-chimeras (which have vegetative tissue from different kinds of plants and which originate by grafting) may also constitute a cultivar.

Seed-produced

Some cultivars "come true from seed", retaining their distinguishing characteristics when grown from seed. Such plants are termed a "variety", "selection" or "strain" but these are ambiguous and confusing words that are best avoided. In general, asexually propagated cultivars grown from seeds produce highly variable seedling plants, and should not be labelled with, or sold under, the parent cultivar's name.

Seed-raised cultivars may be produced by uncontrolled pollination when characteristics that are distinct, uniform and stable are passed from parents to progeny. Some are produced as "lines" that are produced by repeated self-fertilization or inbreeding or "multilines" that are made up of several closely related lines. Sometimes they are F1 hybrids which are the result of a deliberate repeatable single cross between two pure lines. A few F2 hybrid seed cultivars also exist, such as *Achillea* 'Summer Berries'.

Some cultivars are agamospermous plants, which retain their genetic composition and characteristics under reproduction. Occasionally cultivars are raised from seed of a specially selected provenance – for example the seed may be taken from plants that are resistant to a particular disease.

Genetically Modified

Genetically modified plants with characteristics resulting from the deliberate implantation of genetic material from a different germplasm may form a cultivar. However, the International Code of Nomenclature for Cultivated Plants notes, "In practice such an assemblage is often marketed from one or more lines or multilines that have been genetically modified. These lines or multilines often remain in a constant state of development which makes the naming of such an assemblage as a cultivar a futile exercise." However, retired transgenic varieties such as the Fish tomato, which are no longer being developed, do not run into this obstacle and can be given a cultivar name.

Cultivars may be selected because of a change in the ploidy level of a plant which may produce more desirable characteristics.

Cultivar Names

Viola 'Clear Crystals Apricot'
The specific epithet may be omitted from a cultivar name

Every unique cultivar has a unique name within its denomination class (which is almost always the genus). Names of cultivars are regulated by the *International Code of Nomenclature for Cultivated Plants*, and may be registered with an International Cultivar Registration Authority (ICRA).

There are sometimes separate registration authorities for different plant types such as roses and camellias. In addition, cultivars may be associated with commercial marketing names referred to in the *Cultivated Plant Code* as "trade designations".

Presenting in Text

A *cultivar name* consists of a botanical name (of a genus, species, infraspecific taxon, interspecific hybrid or intergeneric hybrid) followed by a cultivar epithet. The cultivar epithet is enclosed by single quotes; it should not be italicized if the botanical name is italicized; and each of the words within the epithet is capitalized (with some permitted exceptions such as conjunctions). It is permissible to place a cultivar epithet after a common name provided the common name is botanically unambiguous. Cultivar epithets published before 1 January 1959 were often given a Latin form and can be readily confused with the specific epithets in botanical names; after that date, newly coined cultivar epithets must be in a modern vernacular language to distinguish them from botanical epithets.

- Examples of correct text presentation:
- *Cryptomeria japonica* 'Elegans'
- *Chamaecyparis lawsoniana* 'Aureomarginata' (pre-1959 name, Latin in form)
- *Chamaecyparis lawsoniana* 'Golden Wonder' (post-1959 name, English language)
- *Pinus densiflora* 'Akebono' (post-1959 name, Japanese language)
- Apple 'Sundown'
- Some incorrect text presentation examples:
- *Cryptomeria japonica* "Elegans"
- *Berberis thunbergii* cv. 'Crimson Pygmy' (this once-common usage is now unacceptable, as it is no longer correct to use "cv." in this context; *Berberis thunbergii* 'Crimson Pygmy' is correct)
- *Rosa* cv. 'Peace' (this is now incorrect for two reasons: firstly, the use of "cv."; secondly, "Peace" is a trade designation or "selling name" for the cultivar *R.* 'Madame A. Meilland' and should therefore be printed in a different typeface from the rest of the name, without quote marks, for example: *Rosa* Peace).

Group Names

Where several very similar cultivars exist they can be associated into a *Group* (formerly *Cultivar-group*). As Group names are used with cultivar names it is necessary to understand their way of presentation. Group names are presented in normal type and the first letter of each word capitalised as for cultivars, but they are not placed in single quotes. When used in a name, the first letter of the word "Group" is itself capitalized.

Presenting in Text

- *Brassica oleracea* Capitata Group (the group of cultivars including all typical cabbages)

- *Brassica oleracea* Botrytis Group (the group of cultivars including all typical cauliflowers)
- *Hydrangea macrophylla* Groupe Hortensia (in French) = *Hydrangea macrophylla* Hortensia Group (in English)
- Where cited with a cultivar name the group should be enclosed in parentheses, as follows:
- *Hydrangea macrophylla* (Hortensia Group) 'Ayesha'

Legal Protection of Cultivars and their Names

Since the 1990s there has been an increasing use of legal protection for newly produced cultivars. Plant breeders expect legal protection for the cultivars they produce. If other growers can immediately propagate and sell these cultivars as soon as they come on the market, the breeder's benefit is largely lost. Legal protection for cultivars is obtained through the use of Plant breeders' rights and plant Patents but the specific legislation and procedures needed to take advantage of this protection vary from country to country.

Controversial use of Legal Protection for Cultivars

The use of legal protection for cultivars can be controversial, particularly for food crops that are staples in developing countries, or for plants selected from the wild and propagated for sale without any additional breeding work; some people consider this practice unethical.

Trade Designations

The formal scientific name of a cultivar, like *Solanum tuberosum* 'King Edward', is a way of uniquely designating a particular kind of plant. This scientific name is in the public domain and cannot be legally protected. Plant retailers wish to maximize their share of the market and one way of doing this is to replace the cumbersome Latin scientific names on plant labels in retail outlets with appealing marketing names that are easy to use, pronounce and remember. Marketing names lie outside the scope of the *Cultivated Plant Code* which refers to them as "trade designations". If a retailer or wholesaler has the sole legal rights to a marketing name then that may offer a sales advantage. Plants protected by Plant breeders' rights (PBR) may have a "true" cultivar name – the recognized scientific name in the public domain, and a "commercial synonym" an additional marketing name that is legally protected: an example would be *Rosa* Fascination = 'Poulmax', the 'Poulmax' being the true scientific name. Because a name that is attractive in one language may have less appeal in another country, a plant may be given different selling names from country to country. Quoting the original cultivar name allows the correct identification of cultivars around the world. The peak body coordinating Plant breeders rights is the Union for the Protection of New Varieties of Plants (*Union internationale pour la protection des obtentions végétales*, UPOV) and this organization maintains a database of new cultivars protected by PBR in all countries.

International Cultivar Registration Authorities

An *International Cultivar Registration Authority* (ICRA) is a voluntary, non-statutory organization appointed by the *Commission for Nomenclature and Cultivar Registration* of the International Society of Horticultural Science. ICRAs are generally formed by societies and institutions specializing in particular plant genera such as *Dahlia* or *Rhododendron* and are currently located in Europe, North America, China, India, Singapore, Australia, New Zealand, South Africa and Puerto Rico.

Dahlia 'Akita' A cultivar selected for flower form and colour

Each ICRA produces an annual report and its reappointment is considered every four years. The main task is to maintain a register of the names within the group of interest and where possible this is published and placed in the public domain. One major aim is to prevent the duplication of cultivar and Group epithets within a genus, as well as ensuring that names are in accord with the latest edition of the *Cultivated Plant Code*. In this way, over the last 50 years or so, ICRAs have contributed to the stability of cultivated plant nomenclature. In recent times many ICRAs have also recorded trade designations and trademarks used in labelling plant material, to avoid confusion with established names.

New names and other relevant data are collected by and submitted to the ICRA and in most cases there is no cost. The ICRA then checks each new epithet to ensure that it has not been used before and that it conforms with the *Cultivated Plant Code*. Each ICRA also ensures that new names are formally established (i.e. published in hard copy, with a description in a dated publication). They record details about the plant, such as parentage, the names of those concerned with its development and introduction, and a basic description highlighting its distinctive characters. ICRAs are not responsible for assessing the distinctiveness of the plant in question. Most ICRAs can be contacted electronically and many maintain web sites: for an up-to-date listing.

Cultigen

A cultigen (from the Latin *cultus* – cultivated, and *gens* – kind) is a plant that has been deliberately altered or selected by humans; it is the result of artificial selection. These "man-made" or anthropogenic plants are, for the most part, plants of commerce that are used in horticulture, agriculture and forestry. Because cultigens are defined by their mode of origin and not by where they are growing, plants meeting this definition remain cultigens whether they are naturalised in the wild, deliberately planted in the wild, or growing in cultivation. Cultigens arise in the following ways: selections of variants from the wild or cultivation including vegetative sports (aberrant growth that can be reproduced reliably in cultivation); plants that are the result of plant breeding and selection programs; genetically modified plants (plants modified by the deliberate implantation of genetic material); and graft-chimaeras (plants grafted to produce mixed tissue, the graft material possibly from wild plants, special selections, or hybrids).

Cultigens may be named in any of a number of ways. The traditional method of scientific naming is under the *International Code of Nomenclature for algae, fungi, and plants*, and many of the most important cultigens, like maize (*Zea mays*) and banana (*Musa acuminata*), are so named. Although it is perfectly in order to give a cultigen a botanical name, in any rank desired, now or at any other time, these days it is more common for cultigens to be given names in accordance with the principles, rules and recommendations laid down in the International Code of Nomenclature for Cultivated Plants (ICNCP) which provides for the names of cultigens in three classification categories, the cultivar, the Group (formerly cultivar-group), and the grex. From that viewpoint it may be said that there is a separate discipline of cultivated plant taxonomy, which forms one of the ways to look at cultigens. The ICNCP does not recognize the use of trade designations and other marketing devices as scientifically acceptable names, but does provide advice on how they should be presented.

Not all cultigens have been given names according to the *Cultivated Plant Code*. Apart from ancient cultigens like those mentioned above there may be occasional anthropogenic plants such as those that are the result of breeding, selection, and tissue grafting that are of no commercial value and have therefore not been given names according to the ICNCP.

Formal Definition

A cultigen is a plant whose origin or selection is primarily due to intentional human activity.

The Wild/cultivated Distinction

Interest in the distinction between wild and cultivated plants dates back to antiquity. Botanical historian Alan Morton notes that wild and cultivated plants (cultigens) were of intense interest to the ancient Greek botanists (partly for religious reasons) and that the distinction was discussed in some detail by Theophrastus (370–285 BCE) the "Father of Botany". Theophrastus was a pupil of both Plato and Aristotle and succeeded the latter as head of the Peripatetic School of Philosophy at the Lyceum in Athens. Theophrastus accepted the view that it was human action, not divine intervention, that produced cultivated plants (cultigens) from wild plants and he also "*had an inkling of the limits of culturally induced (phenotypic) changes and of the importance of genetic constitution*" (*Historia Plantarum* III, 2,2 and *Causa Plantarum* I, 9,3). He also noted that cultivated varieties of fruit trees would degenerate if cultivated from seed.

Origin of Term

The word cultigen was coined in 1918 by Liberty Hyde Bailey (1858–1954) an American horticulturist, botanist and cofounder of the American Society for Horticultural Science. He was aware of the need for special categories for those cultivated plants that had arisen by intentional human activity and which would not fit neatly into the Linnaean hierarchical classification of ranks used by the *International Rules of Botanical Nomenclature* (which later became the *International Code of Nomenclature for algae, fungi, and plants*).

In his 1918 paper Bailey noted that for anyone preparing a descriptive account of the cultivated plants of a country (he was at that time preparing such an account for North America) it would be clear that there are two *gentes* or kinds (Latin singular, *gens*; plural, *gentes*) of plants. Firstly, those that are of known origin or nativity "of known habitat". These he referred to as indigens. The other kind was:

" ... a domesticated group of which the origin may be unknown or indefinite, which has such characters as to separate it from known indigens, and which is probably not represented by any type specimen or exact description, having therefore no clear taxonomic beginning."

He called this second kind of plant a cultigen, the word derived from the conflation of the Latin *cultus* – cultivated, and *gens* – kind.

In 1923 Bailey extended his original discussion emphasising that he was dealing with plants at the rank of species and he referred to indigens as:

- *"those that are discovered in the wild"*

and cultigens as plants that:

- *"arise in some way under the hand of man"*

He then defined a cultigen as:

- *"a species, or its equivalent, that has appeared under domestication"*

Bailey's Definitions

Bailey soon altered his 1923 definition of cultigen when, in 1924, he gave a new definition in the Glossary of his *Manual of Cultivated Plants* as:

" *Plant or group known only in cultivation; presumably originating under domestication; contrast with indigen* "

This, in essence, is the definition given at the head of this piece. This definition of the cultigen permits the recognition of cultivars, unlike the 1923 definition which restricts the idea of the cultigen to plants at the rank of species.

In later publications of the *Liberty Hyde Bailey Hortorium*, Cornell, the idea of the cultigen having the rank of species returned (e.g. *Hortus Second* in 1941 and *Hortus Third* in 1976): both of these publications indicate that the terms cultigen and cultivar are not synonymous and that cultigens exist at the rank of species only.

"*A cultigen is a plant or group of apparent specific rank, known only in cultivation, with no determined nativity, presumably having originated, in the form in which we know it, under domestication. Compare indigen. Examples are* Cucurbita maxima, Phaseolus vulgaris, Zea mays ".

Recent usage in horticulture has, however, maintained a distinction between cultigen and cultivar while nevertheless *allowing the inclusion of cultivars* within the definition.

Cultivars

Cultigen and cultivar may be confused with one another. Cultigen is a general-purpose term encompassing not only plants with cultivar names but others as well, while cultivar is a formal classification category (in the ICNCP).

Although in his 1923 paper Bailey used only the rank of species for the cultigen, it was clear to him that many domesticated plants were more like botanical varieties than species and so he established a new classification category for these, the cultivar, generally assumed to be a contraction of the words "cultivated" and "variety". Bailey was never explicit about the etymology of the word cultivar and it has been suggested that it is a contraction of the words "cultigen" and "variety". He defined cultivar in his 1923 paper as:

... "*a race subordinate to species, that has originated and persisted under cultivation; it is not necessarily, however, referable to a recognised botanical species. It is essentially the equivalent of the botanical variety except in respect to its origin*".

This definition and understanding of cultivar has changed over time.

Usage

Usage in Botany

In botanical literature the word cultigen is generally used to denote a plant which, like the bread wheat (*Triticum aestivum*) is of unknown origin, but presumed to be an ancient human selection. Plants like bread wheat have been given binomials according to the *Botanical Code* and therefore have names with the same form as those of plant species that occur naturally in the wild, but it is not necessary for a cultigen to have a species name, or to have the biological characteristics that distinguish a species. Cultigens can have names at any of various other ranks, including cultivar names, names in the classification categories of grex and group, variety names, forma names, or they may be plants that have been altered by humans (including genetically modified plants) but which have not been given formal names.

Usage in Horticulture

The year 1953 was an important one for cultivated plant taxonomy because this was the date of publication of the first *International Code of Nomenclature for Cultivated Plants* in which Bailey's term cultivar was introduced. It was also the year that the eponymous journal commemorating the work of Bailey (who died in 1954), *Baileya*, was published. In the first volume of *Baileya* taxonomist and colleague of Bailey, George Lawrence, wrote a short article clarifying the distinction between the new term *cultivar* and the *variety*. In the same article he also tried to clarify the critical term *taxon* which had been introduced by German biologist Meyer in the 1920s but had only just been introduced and accepted in botanical circles. This brief article by Lawrence is useful for its insight into the understanding of the meaning of the word cultigen at this time. He opens the article:

In 1918, L.H. Bailey distinguished those plants originating in cultivation from the native plants by designating the former as cultigens and the latter as indigens (indigenous or native to the region). At the same time he proposed the term cultivar to distinguish varieties originating in cultivation from botanical varieties known first in the wild.

In horticulture the definition and use of the terms cultigen and cultivar has varied. One example is the definition given in the Botanical Glossary of *The New Royal Horticultural Society Dictionary of Gardening* which defines cultigen as:

" A plant found only in cultivation or in the wild having escaped from cultivation; included here are many hybrids and cultivars, "...

The use of cultigen in this sense is essentially the same as the definition of the cultigen published by Bailey in 1924.

The *Cultivated Plant Code*, however, states that cultigens are "maintained as recognisable entities solely by continued propagation", and thus would not include plants that have evolved subsequent to escape from cultivation.

Recommended Usage

Wider use of the term cultigen as defined here has been proposed for the following reasons:

- supports current usage in horticulture
- assists clarity in non-technical discussions about "wild" and "cultivated" plants (for example, cultivated plants as commonly understood (plants in cultivation) are not the same as the "cultivated plants" of the *Cultivated Plant Code*, and the distinction between "wild" and "cultivated" habitats is becoming progressively blurred)
- has the potential to simplify the language and definitions used in the Articles and Recommendations of the *Cultivated Plant Code*
- gives greater precision and clarity to the definition of the respective scope, terminology and concepts of the *Botanical Code* and the *Cultivated Plant Code*
- avoids the potential for confusion within the *Cultivated Plant Code* over its scope, that is, whether it is concerned with:
  - where plants are growing (in the wild or in cultivation)
  - how they originated (whether they are the result of intentional human activity or not)
  - whether it simply provides a mechanism for regulating the names of those cultigens requiring special classification categories that are not part of the Linnaean hierarchy of the *Botanical Code* i.e. cultivar and Group names

Critique of Definition

Potential misunderstandings and questions arising from the definition of cultigen given here have been discussed in the literature and are summarised below.

- Natural and *artificial* selection

  The selection process is termed "artificial" when human preferences or influences have a significant effect on the evolution of a particular population or species. Note: artificial selection is a part of the overall selection process – it does not imply that humans are not part of nature, it is simply useful sometimes to distinguish when there has been human influence on selection (as with cultigens).

- What exactly does *altered* mean?

  There are cases that do not seem to comply with the definition. For example, we can presume that the entire global flora is changing as a result of human-induced climate change. Does this mean that all plants are cultigens?

  In cases like this the definition refers to "deliberate" selection and this would be of particular plant characteristics that are not exhibited by a plant's wild counterparts.

- What exactly does *deliberately selected* mean?

  From the moment a plant is taken from the wild it is subject to human selection pressure – from the selection of the original propagation material to the purchase of the plant in a nursery. Surely this form of selection is not deliberate? Again, the early human selection of crops 7,000-10,000 years ago is thought to have occurred quite unintentionally. Variants useful to horticulture often arise spontaneously, they are not deliberate products. Are these cases of unintentional, accidental, or unconscious selection?

  There certainly appear to be cases where origin or selection of a plant is not "deliberate". However, the long term propagation of plants that have some utility, usually economic or ornamental, can hardly be regarded as unintentional and these plants will, almost without exception, have characteristic(s) that distinguish them from their wild counterparts.

- What about plants selected from the wild?

  Plants like *Quercus robur*, Pedunculate or English Oak, *Liquidambar styraciflua*, Sweetgum and *Eucalyptus globulus*, Blue Gum grown in parks and gardens are essentially the same as their wild counterparts and are therefore not cultigens. However, occasionally within natural plant variation there occur characters that are of value to horticulture but of little interest to botany. For example a plant might have flowers of several different colours but these may not have been given formal botanical names. It is customary in horticulture to introduce such variants to commerce and to give them cultivar names. Technically these plants have not been deliberately altered in any way from plants growing (or once growing) in the wild but as they are *deliberately selected* and *named* it seems permissible to refer to them as cultigens. These occurrences are very few. The definition could be (clumsily) extended by mentioning that selection can be for "desirable variation that is not recognised in botanical nomenclature".

- What about gene flow between populations?

  Occasionally cultigens escape from cultivation into the wild where they breed with indigenous plants. Selections may be made from the progeny in the wild and brought back into cultivation where they are used for breeding and the results of the breeding again escape into the wild to breed with indigenous plants. *Lantana* has behaved much like this. The genetic material of a cultigen may become part of the gene pool of a population where, over time, it may be largely or completely swamped. In cases like this what plants are to be called cultigens?

Whether a plant is a cultigen or not does not depend on where it is growing. If it complies with the definition then it is a cultigen. Cases like this have always been difficult for botanical nomenclature. Unnamed progeny in the wild might be given a name like *Lantana* aff. *camara* (aff. = having affinities with) or may remain unnamed. Its cultigenic origin may or may not be recognised by the allocation of a cultivar name.

- Plants of unknown origin

  Occasionally plants will occur whose origin is unknown. Plants growing in cultivation that are unknown in the wild may be determined as cultigenic as a result of scientific investigation, but may remain a mystery.

- Difficult cases

  It may happen that a hybrid cross that has occurred in nature is also performed deliberately in cultivation and that the progeny appear identical. How do we know which plants are cultigens?

  If the cross in cultivation is followed by deliberate selection and naming then this will indicate a cultigen. However in a case like this it may not be possible to tell.

Genetically Modified Crops

Genetically modified crops (GMCs, GM crops, or biotech crops) are plants used in agriculture, the DNA of which has been modified using genetic engineering techniques. In most cases, the aim is to introduce a new trait to the plant which does not occur naturally in the species. Examples in food crops include resistance to certain pests, diseases, or environmental conditions, reduction of spoilage, or resistance to chemical treatments (e.g. resistance to a herbicide), or improving the nutrient profile of the crop. Examples in non-food crops include production of pharmaceutical agents, biofuels, and other industrially useful goods, as well as for bioremediation.

Farmers have widely adopted GM technology. Between 1996 and 2015, the total surface area of land cultivated with GM crops increased by a factor of 100, from 17,000 km^2 (4.2 million acres) to 1,797,000 km^2 (444 million acres). 10% of the world's arable land was planted with GM crops in 2010. In the US, by 2014, 94% of the planted area of soybeans, 96% of cotton and 93% of corn were genetically modified varieties. Use of GM crops expanded rapidly in developing countries, with about 18 million farmers growing 54% of worldwide GM crops by 2013. A 2014 meta-analysis concluded that GM technology adoption had reduced chemical pesticide use by 37%, increased crop yields by 22%, and increased farmer profits by 68%. This reduction in pesticide use has been ecologically beneficial, but benefits may be reduced by overuse. Yield gains and pesticide reductions are larger for insect-resistant crops than for herbicide-tolerant crops. Yield and profit gains are higher in developing countries than in developed countries.

There is a scientific consensus that currently available food derived from GM crops poses no greater risk to human health than conventional food, but that each GM food needs to be tested on a case-by-case basis before introduction. Nonetheless, members of the public are much less likely than scientists to perceive GM foods as safe. The legal and regulatory status of GM foods varies by country, with some nations banning or restricting them, and others permitting them with widely differing degrees of regulation.

However, opponents have objected to GM crops on several grounds, including environmental concerns, whether food produced from GM crops is safe, whether GM crops are needed to address the world's food needs, whether the foods are readily accessible to poor farmers in developing countries, and concerns raised by the fact these organisms are subject to intellectual property law.

Gene Transfer in Nature and Traditional Agriculture

DNA transfers naturally between organisms. Several natural mechanisms allow gene flow across species. These occur in nature on a large scale – for example, it is one mechanism for the development of antibiotic resistance in bacteria. This is facilitated by transposons, retrotransposons, proviruses and other mobile genetic elements that naturally translocate DNA to new loci in a genome. Movement occurs over an evolutionary time scale.

The introduction of foreign germplasm into crops has been achieved by traditional crop breeders by overcoming species barriers. A hybrid cereal grain was created in 1875, by crossing wheat and rye. Since then important traits including dwarfing genes and rust resistance have been introduced. Plant tissue culture and deliberate mutations have enabled humans to alter the makeup of plant genomes.

History

The first genetically modified crop plant was produced in 1982, an antibiotic-resistant tobacco plant. The first field trials occurred in France and the USA in 1986, when tobacco plants were engineered for herbicide resistance. In 1987, Plant Genetic Systems (Ghent, Belgium), founded by Marc Van Montagu and Jeff Schell, was the first company to genetically engineer insect-resistant (tobacco) plants by incorporating genes that produced insecticidal proteins from *Bacillus thuringiensis* (Bt).

The People's Republic of China was the first country to allow commercialized transgenic plants, introducing a virus-resistant tobacco in 1992, which was withdrawn in 1997. The first genetically modified crop approved for sale in the U.S., in 1994, was the *FlavrSavr* tomato. It had a longer shelf life, because it took longer to soften after ripening. In 1994, the European Union approved tobacco engineered to be resistant to the herbicide bromoxynil, making it the first commercially genetically engineered crop marketed in Europe.

In 1995, Bt Potato was approved by the US Environmental Protection Agency, making it the country's first pesticide producing crop. In 1995 canola with modified oil composition (Calgene), Bt maize (Ciba-Geigy), bromoxynil-resistant cotton (Calgene), Bt cotton (Monsanto), glyphosate-resistant soybeans (Monsanto), virus-resistant squash (Asgrow), and additional delayed ripening tomatoes (DNAP, Zeneca/Peto, and Monsanto) were approved. As of mid-1996, a total of 35 approvals had been granted to commercially grow 8 transgenic crops and one flower crop (carnation), with 8 different traits in 6 countries plus the EU. In 2000, Vitamin A-enriched golden rice was developed, though as of 2016 it was not yet in commercial production. In 2013 the leaders of the three research teams that first applied genetic engineering to crops, Robert Fraley, Marc Van Montagu and Mary-Dell Chilton were awarded the World Food Prize for improving the "quality, quantity or availability" of food in the world.

Methods

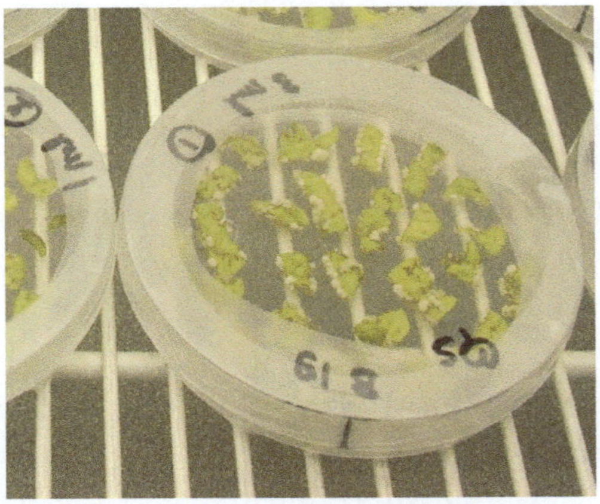

Plants (*Solanum chacoense*) being transformed using agrobacterium

Genetically engineered crops have genes added or removed using genetic engineering techniques, originally including gene guns, electroporation, microinjection and agrobacterium. More recently, CRISPR and TALEN offered much more precise and convenient editing techniques.

Gene guns (also known as biolistics) "shoot" (direct high energy particles or radiations against) target genes into plant cells. It is the most common method. DNA is bound to tiny particles of gold or tungsten which are subsequently shot into plant tissue or single plant cells under high pressure. The accelerated particles penetrate both the cell wall and membranes. The DNA separates from the metal and is integrated into plant DNA inside the nucleus. This method has been applied successfully for many cultivated crops, especially monocots like wheat or maize, for which transformation using *Agrobacterium tumefaciens* has been less successful. The major disadvantage of this procedure is that serious damage can be done to the cellular tissue.

Agrobacterium tumefaciens-mediated transformation is another common technique. Agrobacteria are natural plant parasites, and their natural ability to transfer genes provides another engineering method. To create a suitable environment for themselves, these Agrobacteria insert their genes into plant hosts, resulting in a proliferation of modified plant cells near the soil level (crown gall). The genetic information for tumor growth is encoded on a mobile, circular DNA fragment (plasmid). When *Agrobacterium* infects a plant, it transfers this T-DNA to a random site in the plant genome. When used in genetic engineering the bacterial T-DNA is removed from the bacterial plasmid and replaced with the desired foreign gene. The bacterium is a vector, enabling transportation of foreign genes into plants. This method works especially well for dicotyledonous plants like potatoes, tomatoes, and tobacco. Agrobacteria infection is less successful in crops like wheat and maize.

Electroporation is used when the plant tissue does not contain cell walls. In this technique, "DNA enters the plant cells through miniature pores which are temporarily caused by electric pulses."

Microinjection directly injects the gene into the DNA.

Plant scientists, backed by results of modern comprehensive profiling of crop composition, point out that crops modified using GM techniques are less likely to have unintended changes than are conventionally bred crops.

In research tobacco and *Arabidopsis thaliana* are the most frequently modified plants, due to well-developed transformation methods, easy propagation and well studied genomes. They serve as model organisms for other plant species.

Introducing new genes into plants requires a promoter specific to the area where the gene is to be expressed. For instance, to express a gene only in rice grains and not in leaves, an endosperm-specific promoter is used. The codons of the gene must be optimized for the organism due to codon usage bias.

Types of Modifications

Transgenic maize containing a gene from the bacteria *Bacillus thuringiensis*

Transgenic

Transgenic plants have genes inserted into them that are derived from another species. The inserted genes can come from species within the same kingdom (plant to plant) or between kingdoms (for example, bacteria to plant). In many cases the inserted DNA has to be modified slightly in order to correctly and efficiently express in the host organism. Transgenic plants are used to express proteins like the cry toxins from *B. thuringiensis*, herbicide resistant genes, antibodies and antigens for vaccinations A study led by the European Food Safety Authority (EFSA) found also viral genes in transgenic plants.

Transgenic carrots have been used to produce the drug Taliglucerase alfa which is used to treat Gaucher's disease. In the laboratory, transgenic plants have been modified to increase photosynthesis (currently about 2% at most plants versus the theoretic potential of 9–10%). This is possible by changing the rubisco enzyme (i.e. changing C3 plants into C4 plants), by placing the rubisco in a carboxysome, by adding CO_2 pumps in the cell wall, by changing the leaf form/size. Plants have been engineered to exhibit bioluminescence that may become a sustainable alternative to electric lighting.

Cisgenic

Cisgenic plants are made using genes found within the same species or a closely related one, where conventional plant breeding can occur. Some breeders and scientists argue that cisgenic modification is useful for plants that are difficult to crossbreed by conventional means (such as potatoes), and that plants in the cisgenic category should not require the same regulatory scrutiny as transgenics.

Subgenic

Genetically modified plants can also be developed using gene knockdown or gene knockout to alter the genetic makeup of a plant without incorporating genes from other plants. In 2014, Chinese researcher Gao Caixia filed patents on the creation of a strain of wheat that is resistant to powdery mildew. The strain lacks genes that encode proteins that repress defenses against the mildew. The researchers deleted all three copies of the genes from wheat's hexaploid genome. Gao used the TALENs and CRISPR gene editing tools without adding or changing any other genes. No field trials were immediately planned. The CRISPR technique has also been used to modify white button mushrooms (*Agaricus bisporus*).

Economics

GM food's economic value to farmers is one of its major benefits, including in developing nations. A 2010 study found that Bt corn provided economic benefits of $6.9 billion over the previous 14 years in five Midwestern states. The majority ($4.3 billion) accrued to farmers producing non-Bt corn. This was attributed to European corn borer populations reduced by exposure to Bt corn, leaving fewer to attack conventional corn nearby. Agriculture economists calculated that "world surplus [increased by] $240.3 million for 1996. Of this total, the largest share (59%) went to U.S. farmers. Seed company Monsanto received the next largest share (21%), followed by US consumers (9%), the rest of the world (6%), and the germplasm supplier, Delta & Pine Land Company of Mississippi (5%)."

According to the International Service for the Acquisition of Agri-biotech Applications (ISAAA), in 2014 approximately 18 million farmers grew biotech crops in 28 countries; about 94% of the farmers were resource-poor in developing countries. 53% of the global biotech crop area of 181.5 million hectares was grown in 20 developing countries. PG Economics comprehensive 2012 study concluded that GM crops increased farm incomes worldwide by $14 billion in 2010, with over half this total going to farmers in developing countries.

Critics challenged the claimed benefits to farmers over the prevalence of biased observers and by the absence of randomized controlled trials. The main Bt crop grown by small farmers in developing countries is cotton. A 2006 review of Bt cotton findings by agricultural economists concluded, "the overall balance sheet, though promising, is mixed. Economic returns are highly variable over years, farm type, and geographical location".

In 2013 the European Academies Science Advisory Council (EASAC) asked the EU to allow the development of agricultural GM technologies to enable more sustainable agriculture, by employing fewer land, water, and nutrient resources. EASAC also criticizes the EU's "time-consuming and expensive regulatory framework" and said that the EU had fallen behind in the adoption of GM technologies.

Participants in agriculture business markets include seed companies, agrochemical companies, distributors, farmers, grain elevators and universities that develop new crops/traits and whose agricultural extensions advise farmers on best practices. According to a 2012 review based on data from the late 1990s and early 2000s, much of the GM crop grown each year is used for livestock feed and increased demand for meat leads to increased demand for GM feed crops. Feed grain usage as a percentage of total crop production is 70% for corn and more than 90% of oil seed meals such as soybeans. About 65 million metric tons of GM corn grains and about 70 million metric tons of soybean meals derived from GM soybean become feed.

In 2014 the global value of biotech seed was US$15.7 billion; US$11.3 billion (72%) was in industrial countries and US$4.4 billion (28%) was in the developing countries. In 2009, Monsanto had $7.3 billion in sales of seeds and from licensing its technology; DuPont, through its Pioneer subsidiary, was the next biggest company in that market. As of 2009, the overall Roundup line of products including the GM seeds represented about 50% of Monsanto's business.

Some patents on GM traits have expired, allowing the legal development of generic strains that include these traits. For example, generic glyphosate-tolerant GM soybean is now available. Another impact is that traits developed by one vendor can be added to another vendor's proprietary strains, potentially increasing product choice and competition. The patent on the first type of *Roundup Ready* crop that Monsanto produced (soybeans) expired in 2014 and the first harvest of off-patent soybeans occurs in the spring of 2015. Monsanto has broadly licensed the patent to other seed companies that include the glyphosate resistance trait in their seed products. About 150 companies have licensed the technology, including Syngenta and DuPont Pioneer.

Yield

In 2014, the largest review yet concluded that GM crops' effects on farming were positive. The meta-analysis considered all published English-language examinations of the agronomic and economic impacts between 1995 and March 2014 for three major GM crops: soybean, maize, and cotton. The study found that herbicide-tolerant crops have lower production costs, while for insect-resistant crops the reduced pesticide use was offset by higher seed prices, leaving overall production costs about the same.

Yields increased 9% for herbicide tolerance and 25% for insect resistant varieties. Farmers who adopted GM crops made 69% higher profits than those who did not. The review found that GM crops help farmers in developing countries, increasing yields by 14 percentage points.

The researchers considered some studies that were not peer-reviewed and a few that did not report sample sizes. They attempted to correct for publication bias, by considering sources beyond academic journals. The large data set allowed the study to control for potentially confounding variables such as fertilizer use. Separately, they concluded that the funding source did not influence study results.

Traits

GM crops grown today, or under development, have been modified with various traits. These traits include improved shelf life, disease resistance, stress resistance, herbicide resistance, pest resistance, production of useful goods such as biofuel or drugs, and ability to absorb toxins and for use in bioremediation of pollution.

Recently, research and development has been targeted to enhancement of crops that are locally important in developing countries, such as insect-resistant cowpea for Africa and insect-resistant brinjal (eggplant).

Lifetime

The first genetically modified crop approved for sale in the U.S. was the *FlavrSavr* tomato, which had a longer shelf life. It is no longer on the market.

In November 2014, the USDA approved a GM potato that prevents bruising.

In February 2015 Arctic Apples were approved by the USDA, becoming the first genetically modified apple approved for US sale. Gene silencing was used to reduce the expression of polyphenol oxidase (PPO), thus preventing enzymatic browning of the fruit after it has been sliced open. The trait was added to Granny Smith and Golden Delicious varieties. The trait includes a bacterial antibiotic resistance gene that provides resistance to the antibiotic kanamycin. The genetic engineering involved cultivation in the presence of kanamycin, which allowed only resistant cultivars to survive. Humans consuming apples do not acquire kanamycin resistance, per arcticapple.com. The FDA approved the apples in March 2015.

Nutrition

Edible Oils

Some GM soybeans offer improved oil profiles for processing or healthier eating. Camelina sativa has been modified to produce plants that accumulate high levels of oils similar to fish oils.

Vitamin Enrichment

Golden rice, developed by the International Rice Research Institute (IRRI), provides greater amounts of Vitamin A targeted at reducing Vitamin A deficiency. As of January 2016, golden rice has not yet been grown commercially in any country.

Researchers vitamin-enriched corn derived from South African white corn variety M37W, producing a 169-fold increase in Vitamin A, 6-fold increase in Vitamin C and doubled concentrations of folate. Modified Cavendish bananas express 10-fold the amount of Vitamin A as unmodified varieties.

Toxin Reduction

A genetically modified cassava under development offers lower cyanogen glucosides and enhanced protein and other nutrients (called BioCassava).

In November 2014, the USDA approved a potato, developed by J.R. Simplot Company, that prevents bruising and produces less acrylamide when fried. The modifications prevent natural, harmful proteins from being made via RNA interference. They do not employ genes from non-potato species. The trait was added to the Russet Burbank, Ranger Russet and Atlantic varieties.

Stress Resistance

Plants engineered to tolerate non-biological stressors such as drought, frost, high soil salinity, and nitrogen starvation were in development. In 2011, Monsanto's DroughtGard maize became the first drought-resistant GM crop to receive US marketing approval.

Herbicides

Glyphosate

As of 1999 the most prevalent GM trait was glyphosate-resistance. Glyphosate, (the active ingredient in Roundup and other herbicide products) kills plants by interfering with the shikimate pathway in plants, which is essential for the synthesis of the aromatic amino acids phenylalanine, tyrosine and tryptophan. The shikimate pathway is not present in animals, which instead obtain aromatic amino acids from their diet. More specifically, glyphosate inhibits the enzyme 5-enolpyruvylshikimate-3-phosphate synthase (EPSPS).

This trait was developed because the herbicides used on grain and grass crops at the time were highly toxic and not effective against narrow-leaved weeds. Thus, developing crops that could withstand spraying with glyphosate would both reduce environmental and health risks, and give an agricultural edge to the farmer.

Some micro-organisms have a version of EPSPS that is resistant to glyphosate inhibition. One of these was isolated from an *Agrobacterium* strain CP4 (CP4 EPSPS) that was resistant to glyphosate. The CP4 EPSPS gene was engineered for plant expression by fusing the 5' end of the gene to a chloroplast transit peptide derived from the petunia EPSPS. This transit peptide was used because it had shown previously an ability to deliver bacterial EPSPS to the chloroplasts of other plants. This CP4 EPSPS gene was cloned and transfected into soybeans.

The plasmid used to move the gene into soybeans was PV-GMGTO4. It contained three bacterial genes, two CP4 EPSPS genes, and a gene encoding beta-glucuronidase (GUS) from *Escherichia coli* as a marker. The DNA was injected into the soybeans using the particle acceleration method. Soybean cultivar A5403 was used for the transformation.

Bromoxynil

Tobacco plants have been engineered to be resistant to the herbicide bromoxynil.

Glufosinate

Crops have been commercialized that are resistant to the herbicide glufosinate, as well. Crops engineered for resistance to multiple herbicides to allow farmers to use a mixed group of two, three, or four different chemicals are under development to combat growing herbicide resistance.

2,4-D

In October 2014 the US EPA registered Dow's Enlist Duo maize, which is genetically modified to be resistant to both glyphosate and 2,4-D, in six states. Inserting a bacterial aryloxyalkanoate dioxygenase gene, *aad1* makes the corn resistant to 2,4-D. The USDA had approved maize and soybeans with the mutation in September 2014.

Dicamba

Monsanto has requested approval for a stacked strain that is tolerant of both glyphosate and dicamba.

Pest Resistance

Insects

Tobacco, corn, rice and many other crops have been engineered to express genes encoding for insecticidal proteins from Bacillus thuringiensis (Bt). Papaya, potatoes, and squash have been engineered to resist viral pathogens such as cucumber mosaic virus which, despite its name, infects a wide variety of plants. The introduction of Bt crops during the period between 1996 and 2005 has been estimated to have reduced the total volume of insecticide active ingredient use in the United States by over 100 thousand tons. This represents a 19.4% reduction in insecticide use.

In the late 1990s, a genetically modified potato that was resistant to the Colorado potato beetle was withdrawn because major buyers rejected it, fearing consumer opposition.

Viruses

Virus resistant papaya were developed in response to a papaya ringspot virus (PRV) outbreak in Hawaii in the late 1990s. They incorporate PRV DNA. By 2010, 80% of Hawaiian papaya plants were genetically modified.

Potatoes were engineered for resistance to potato leaf roll virus and Potato virus Y in 1998. Poor sales led to their market withdrawal after three years.

Yellow squash that were resistant to at first two, then three viruses were developed, beginning in the 1990s. The viruses are watermelon, cucumber and zucchini/courgette yellow mosaic. Squash was the second GM crop to be approved by US regulators. The trait was later added to zucchini.

Many strains of corn have been developed in recent years to combat the spread of Maize dwarf mosaic virus, a costly virus that causes stunted growth which is carried in Johnson grass and spread by aphid insect vectors. These strands are commercially available although the resistance is not standard among GM corn variants.

By-products

Drugs

In 2012, the FDA approved the first plant-produced pharmaceutical, a treatment for Gaucher's Disease. Tobacco plants have been modified to produce therapeutic antibodies.

Biofuel

Algae is under development for use in biofuels. Researchers in Singapore were working on GM jatropha for biofuel production. Syngenta has USDA approval to market a maize trademarked Enogen that has been genetically modified to convert its starch to sugar for ethanol. In 2013, the Flemish Institute for Biotechnology was investigating poplar trees genetically engineered to contain less lignin to ease conversion into ethanol. Lignin is the critical limiting factor when using wood to make bio-ethanol because lignin limits the accessibility of cellulose microfibrils to depolymerization by enzymes.

Materials

Companies and labs are working on plants that can be used to make bioplastics. Potatoes that produce industrially useful starches have been developed as well. Oilseed can be modified to produce fatty acids for detergents, substitute fuels and petrochemicals.

Bioremediation

Scientists at the University of York developed a weed (*Arabidopsis thaliana*) that contains genes from bacteria that could clean TNT and RDX-explosive soil contaminants in 2011. 16 million hectares in the USA (1.5% of the total surface) are estimated to be contaminated with TNT and RDX. However *A. thaliana* was not tough enough for use on military test grounds. Modifications in 2016 included switchgrass and bentgrass.

Genetically modified plants have been used for bioremediation of contaminated soils. Mercury, selenium and organic pollutants such as polychlorinated biphenyls (PCBs).

Marine environments are especially vulnerable since pollution such as oil spills are not containable. In addition to anthropogenic pollution, millions of tons of petroleum annually enter the marine environment from natural seepages. Despite its toxicity, a considerable fraction of petroleum oil entering marine systems is eliminated by the hydrocarbon-degrading activities of microbial communities. Particularly successful is a recently discovered group of specialists, the so-called hydrocarbonoclastic bacteria (HCCB) that may offer useful genes.

Asexual Reproduction

Crops such as maize reproduce sexually each year. This randomizes which genes get propagated to the next generation, meaning that desirable traits can be lost. To maintain a high-quality crop, some farmers purchase seeds every year. Typically, the seed company maintains two inbred varieties and crosses them into a hybrid strain that is then sold. Related plants like sorghum and gamma grass are able to perform apomixis, a form of asexual reproduction that keeps the plant's DNA intact. This trait is apparently controlled by a single dominant gene, but traditional breeding has been unsuccessful in creating asexually-reproducing maize. Genetic engineering offers another route to this goal. Successful modification would allow farmers to replant harvested seeds that retain desirable traits, rather than relying on purchased seed.

Crops

Herbicide Tolerance

GMO	Use
Alfalfa	Animal feed
Canola	Cooking oil Margarine Emulsifiers in packaged foods
Cotton	Fiber Cottonseed oil Animal feed
Maize	Animal feed high-fructose corn syrup corn starch
Soybean	Animal feed Soybean oil
Sugar Beet	Food

Insect Resistance

GMO	Use
Cotton	Fiber Cottonseed oil Animal feed
Eggplant	Food
Maize	Animal feed high-fructose corn syrup corn starch
Poplar	Tree

Development

The number of USDA-approved field releases for testing grew from 4 in 1985 to 1,194 in 2002 and averaged around 800 per year thereafter. The number of sites per release and the number of gene constructs (ways that the gene of interest is packaged together with other elements)—have rapidly increased since 2005. Releases with agronomic properties (such as drought resistance) jumped from 1,043 in 2005 to 5,190 in 2013. As of September 2013, about 7,800 releases had been approved for corn, more than 2,200 for soybeans, more than 1,100 for cotton, and about 900 for potatoes. Releases were approved for herbicide tolerance (6,772 releases), insect resistance (4,809), product quality such as flavor or nutrition (4,896), agronomic properties like drought resistance (5,190), and virus/fungal resistance (2,616). The institutions with the most authorized field releases include Monsanto with 6,782, Pioneer/DuPont with 1,405, Syngenta with 565, and USDA's Agricultural Research Service with 370. As of September 2013 USDA had received proposals for releasing GM rice, squash, plum, rose, tobacco, flax, and chicory.

Other Modified Traits

GMO	Use
Canola	Cooking oil Margarine Emulsifiers in packaged foods
Carnation	Ornamental
Maize	Animal feed high-fructose corn syrup corn starch

Papaya	Food
Petunia	Ornamental
Potato	Food
	Industrial
Rose	Ornamental
Soybean	Animal feed Soybean oil
Squash	Food
Sugar Cane	Food
Tobacco	Cigarettes

Farming Practices

Large infestations of agricultural pest and herbs increased the usage of GM products. The usage of GM products has made farming less dependent on the farmer because they have the luxury of being assured a clean cro pdue to GM prodcuts.

Resistance

Bt Resistance

Constant exposure to a toxin creates evolutionary pressure for pests resistant to that toxin. Over-reliance on glyphosate and a reduction in the diversity of weed management practices allowed the spread of glyphosate resistance in 14 weed species/biotypes in the US.

One method of reducing resistance is the creation of refuges to allow nonresistant organisms to survive and maintain a susceptible population.

To reduce resistance to Bt crops, the 1996 commercialization of transgenic cotton and maize came with a management strategy to prevent insects from becoming resistant. Insect resistance management plans are mandatory for Bt crops. The aim is to encourage a large population of pests so that any (recessive) resistance genes are diluted within the population. Resistance lowers evolutionary fitness in the absence of the stressor (Bt). In refuges, non-resistant strains outcompete resistant ones.

With sufficiently high levels of transgene expression, nearly all of the heterozygotes (S/s), i.e., the largest segment of the pest population carrying a resistance allele, will be killed before maturation, thus preventing transmission of the resistance gene to their progeny. Refuges (i. e., fields of non-transgenic plants) adjacent to transgenic fields increases the likelihood that homozygous resistant (s/s) individuals and any surviving heterozygotes will mate with susceptible (S/S) individuals from the refuge, instead of with other individuals carrying the resistance allele. As a result, the resistance gene frequency in the population remains lower.

Complicating factors can affect the success of the high-dose/refuge strategy. For example, if the temperature is not ideal, thermal stress can lower Bt toxin production and leave the plant more susceptible. More importantly, reduced late-season expression has been documented, possibly resulting from DNA methylation of the promoter. The success of the high-dose/refuge strategy has successfully maintained the value of Bt crops. This success has depended on factors independent of management strategy, including low initial resistance allele frequencies, fitness costs associated with resistance, and the abundance of non-Bt host plants outside the refuges.

Companies that produce Bt seed are introducing strains with multiple Bt proteins. Monsanto did this with Bt cotton in India, where the product was rapidly adopted. Monsanto has also; in an attempt to simplify the process of implementing refuges in fields to comply with Insect Resistance Management(IRM) policies and prevent irresponsible planting practices; begun marketing seed bags with a set proportion of refuge (non-transgenic) seeds mixed in with the Bt seeds being sold. Coined "Refuge-In-a-Bag" (RIB), this practice is intended to increase farmer compliance with refuge requirements and reduce additional labor needed at planting from having separate Bt and refuge seed bags on hand. This strategy is likely to reduce the likelihood of Bt-resistance occurring for corn rootworm, but may increase the risk of resistance for lepidopteran corn pests, such as European corn borer. Increased concerns for resistance with seed mixtures include partially resistant larvae on a Bt plant being able to move to a susceptible plant to survive or cross pollination of refuge pollen on to Bt plants that can lower the amount of Bt expressed in kernels for ear feeding insects.

Herbicide Resistance

Best management practices (BMPs) to control weeds may help delay resistance. BMPs include applying multiple herbicides with different modes of action, rotating crops, planting weed-free seed, scouting fields routinely, cleaning equipment to reduce the transmission of weeds to other fields, and maintaining field borders. The most widely planted GMOs are designed to tolerate herbicides. By 2006 some weed populations had evolved to tolerate some of the same herbicides. Palmer amaranth is a weed that competes with cotton. A native of the southwestern US, it traveled east and was first found resistant to glyphosate in 2006, less than 10 years after GM cotton was introduced.

Plant Protection

Farmers generally use less insecticide when they plant Bt-resistant crops. Insecticide use on corn farms declined from 0.21 pound per planted acre in 1995 to 0.02 pound in 2010. This is consistent with the decline in European corn borer populations as a direct result of Bt corn and cotton. The establishment of minimum refuge requirements helped delay the evolution of Bt resistance. However, resistance appears to be developing to some Bt traits in some areas.

Tillage

By leaving at least 30% of crop residue on the soil surface from harvest through planting, conservation tillage reduces soil erosion from wind and water, increases water retention, and reduces soil degradation as well as water and chemical runoff. In addition, conservation tillage reduces the carbon footprint of agriculture. A 2014 review covering 12 states from 1996 to 2006, found that a 1% increase in herbicde-tolerant (HT) soybean adoption leads to a 0.21% increase in conservation tillage and a 0.3% decrease in quality-adjusted herbicide use.

Regulation

The regulation of genetic engineering concerns the approaches taken by governments to assess and manage the risks associated with the development and release of genetically modified crops. There are differences in the regulation of GM crops between countries, with some of the most marked differences occurring between the USA and Europe. Regulation varies in a given country depending on the intended use of each product. For example, a crop not intended for food use is generally not reviewed by authorities responsible for food safety.

Production

In 2013, GM crops were planted in 27 countries; 19 were developing countries and 8 were developed countries. 2013 was the second year in which developing countries grew a majority (54%) of the total GM harvest. 18 million farmers grew GM crops; around 90% were small-holding farmers in developing countries.

Country	2013– GM planted area (million hectares)	Biotech crops
USA	70.1	Maize, Soybean, Cotton, Canola, Sugarbeet, Alfalfa, Papaya, Squash
Brazil	40.3	Soybean, Maize, Cotton
Argentina	24.4	Soybean, Maize, Cotton
India	11.0	Cotton
Canada	10.8	Canola, Maize, Soybean, Sugarbeet
Total	175.2	----

The United States Department of Agriculture (USDA) reports every year on the total area of GMO varieties planted in the United States. According to National Agricultural Statistics Service, the states published in these tables represent 81–86 percent of all corn planted area, 88–90 percent of all soybean planted area, and 81–93 percent of all upland cotton planted area (depending on the year).

Global estimates are produced by the International Service for the Acquisition of Agri-biotech Applications (ISAAA) and can be found in their annual reports, "Global Status of Commercialized Transgenic Crops".

Farmers have widely adopted GM technology. Between 1996 and 2013, the total surface area of land cultivated with GM crops increased by a factor of 100, from 17,000 square kilometers (4,200,000 acres) to 1,750,000 km² (432 million acres). 10% of the world's arable land was planted with GM crops in 2010. As of 2011, 11 different transgenic crops were grown commercially on 395 million acres (160 million hectares) in 29 countries such as the USA, Brazil, Argentina, India, Canada, China, Paraguay, Pakistan, South Africa, Uruguay, Bolivia, Australia, Philippines, Myanmar, Burkina Faso, Mexico and Spain. One of the key reasons for this widespread adoption is the perceived economic benefit the technology brings to farmers. For example, the system of planting glyphosate-resistant seed and then applying glyphosate once plants emerged provided farmers with the opportunity to dramatically increase the yield from a given plot of land, since this allowed them to plant rows closer together. Without it, farmers had to plant rows far enough apart to control post-emergent weeds with mechanical tillage. Likewise, using Bt seeds means that farmers do not have to purchase insecticides, and then invest time, fuel, and equipment in applying them. However critics have disputed whether yields are higher and whether chemical use is less, with GM crops.

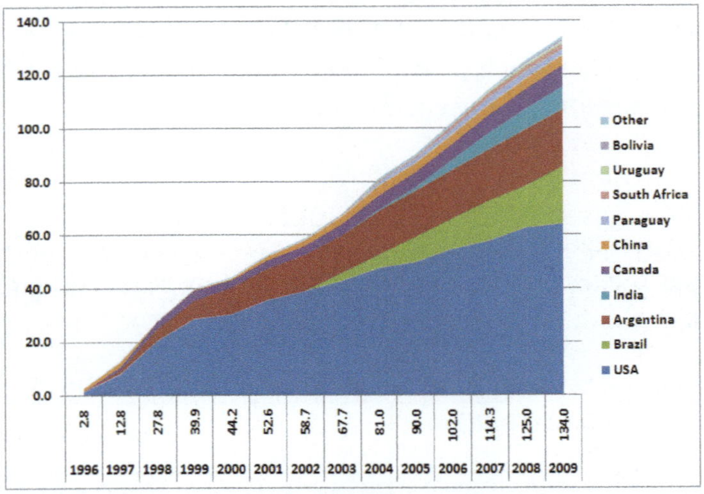

Land area used for genetically modified crops by country (1996–2009), in millions of hectares. In 2011, the land area used was 160 million hectares, or 1.6 million square kilometers.

In the US, by 2014, 94% of the planted area of soybeans, 96% of cotton and 93% of corn were genetically modified varieties. Genetically modified soybeans carried herbicide-tolerant traits only, but maize and cotton carried both herbicide tolerance and insect protection traits (the latter largely Bt protein). These constitute "input-traits" that are aimed to financially benefit the producers, but may have indirect environmental benefits and cost benefits to consumers. The Grocery Manufacturers of America estimated in 2003 that 70–75% of all processed foods in the U.S. contained a GM ingredient.

Europe grows relatively few genetically engineered crops with the exception of Spain, where one fifth of maize is genetically engineered, and smaller amounts in five other countries. The EU had a 'de facto' ban on the approval of new GM crops, from 1999 until 2004. GM crops are now regulated by the EU. In 2015, genetically engineered crops are banned in 38 countries worldwide, 19 of them in Europe. Developing countries grew 54 percent of genetically engineered crops in 2013.

In recent years GM crops expanded rapidly in developing countries. In 2013 approximately 18 million farmers grew 54% of worldwide GM crops in developing countries. 2013's largest increase was in Brazil (403,000 km^2 versus 368,000 km^2 in 2012). GM cotton began growing in India in 2002, reaching 110,000 km^2 in 2013.

According to the 2013 ISAAA brief: "...a total of 36 countries (35 + EU-28) have granted regulatory approvals for biotech crops for food and/or feed use and for environmental release or planting since 1994... a total of 2,833 regulatory approvals involving 27 GM crops and 336 GM events (NB: an "event" is a specific genetic modification in a specific species) have been issued by authorities, of which 1,321 are for food use (direct use or processing), 918 for feed use (direct use or processing) and 599 for environmental release or planting. Japan has the largest number (198), followed by the U.S.A. (165, not including "stacked" events), Canada (146), Mexico (131), South Korea (103),

Australia (93), New Zealand (83), European Union (71 including approvals that have expired or under renewal process), Philippines (68), Taiwan (65), Colombia (59), China (55) and South Africa (52). Maize has the largest number (130 events in 27 countries), followed by cotton (49 events in 22 countries), potato (31 events in 10 countries), canola (30 events in 12 countries) and soybean (27 events in 26 countries).

Controversy

GM foods are controversial and the subject of protests, vandalism, referenda, legislation, court action and scientific disputes. The controversies involve consumers, biotechnology companies, governmental regulators, non-governmental organizations and scientists. The key areas are whether GM food should be labeled, the role of government regulators, the effect of GM crops on health and the environment, the effects of pesticide use and resistance, the impact on farmers, and their roles in feeding the world and energy production.

There is a scientific consensus that currently available food derived from GM crops poses no greater risk to human health than conventional food, but that each GM food needs to be tested on a case-by-case basis before introduction. Nonetheless, members of the public are much less likely than scientists to perceive GM foods as safe. The legal and regulatory status of GM foods varies by country, with some nations banning or restricting them, and others permitting them with widely differing degrees of regulation.

No reports of ill effects have been documented in the human population from GM food. Although GMO labeling is required in many countries, the United States Food and Drug Administration does not require labeling, nor does it recognize a distinction between approved GMO and non-GMO foods.

Advocacy groups such as Center for Food Safety, Union of Concerned Scientists, Greenpeace and the World Wildlife Fund claim that risks related to GM food have not been adequately examined and managed, that GMOs are not sufficiently tested and should be labelled, and that regulatory authorities and scientific bodies are too closely tied to industry. Some studies have claimed that genetically modified crops can cause harm; a 2016 review that reanalyzed the data from six of these studies found that their statistical methodologies were flawed and did not demonstrate harm, and said that conclusions about GMO crop safety should be drawn from "the totality of the evidence... instead of far-fetched evidence from single studies".

References

- Andrew Pollack (19 June 2013). "Executive at Monsanto wins global food honor". The New York Times. Retrieved 20 June 2013
- Marris, Claire (2001). "Public views on GMOs: deconstructing the myths" (PDF). EMBO Reports. 2 (7): 545–548. doi:10.1093/embo-reports/kve142. PMC 1083956. PMID 11463731
- Davis, D.R.; Epp, M.D.; Riordan, H.D. (2004). "Changes in USDA Food Composition Data for 43 Garden Crops, 1950 to 1999". Journal of the American College of Nutrition. 23 (6): 669–682. doi:10.1080/07315724.2004.10719409
- Duncan, R. (1996). "Tissue Culture-Induced Variation and Crop Improvement". Advances in Agronomy. Advances in Agronomy. 58: 201–240. doi:10.1016/S0065-2113(08)60256-4. ISBN 9780120007585

- Bänziger (2000). "Breeding for drought and nitrogen stress tolerance in maize: from theory to practice". From Theory to Practice: 7–9. Retrieved 7-11-2013
- Scott, Sydney E.; Inbar, Yoel; Rozin, Paul (2016). "Evidence for Absolute Moral Opposition to Genetically Modified Food in the United States" (PDF). Perspectives on Psychological Science. 11 (3): 315–324. doi:10.1177/1745691615621275. PMID 27217243
- Korbin, M. U. (2011). Molecular approaches to disease resistance in fragaria spp. Journal of Plant Protection Research, 51(1), 60-65
- Martins VAP (2008). "Genomic Insights into Oil Biodegradation in Marine Systems". Microbial Biodegradation: Genomics and Molecular Biology. Caister Academic Press. ISBN 978-1-904455-17-2
- Brody, J.E. (December 5, 2000). "Personal Health; Gene Altered Foods: A Case Against Panic". New York Times. Retrieved 11/8/2011
- Lederberg J, Tatum EL; Tatum (1946). "Gene recombination in E. coli". Nature. 158 (4016): 558. Bibcode:-1946Natur.158..558L. doi:10.1038/158558a0
- Monroe D. (2006). "Jumping Genes Cross Plant Species Boundaries". PLoS Biology. 4 (1): e35. doi:10.1371/journal.pbio.0040035
- ISAAA 2013 Annual Report Executive Summary, Global Status of Commercialized Biotech/GM Crops: 2013 ISAAA Brief 46-2013, Retrieved 6 August 2014
- Costa-Font, J.; Mossialos, E. (2007). "Are perceptions of 'risks' and 'benefits' of genetically modified food (in) dependent?". Food Quality and Preference. 18: 173–182. doi:10.1016/j.foodqual.2005.09.013
- Wang, Guo-ying (2009). "Genetic Engineering for Maize Improvement in China". Electronic Journal of Biotechnology. Electronic Journal of Biotechnology. Retrieved December 1, 2015
- Bashshur, Ramona (February 2013). "FDA and Regulation of GMOs". American Bar Association. Retrieved February 24, 2016
- Siebert, S; Döll, P; Hoogeveen, J; Faures, J M; Frenken, K; Feick, S (2005). "Development and validation of the global map of irrigation areas". Hydrol. Earth Syst. Sci. 9: 535–47. doi:10.5194/hess-9-535-2005
- Maxmen, Amy (2 May 2012) First plant-made drug on the market Nature, Biology & Biotechnology, Industry. Retrieved 26 June 2012
- Morgante, M.; Brunner, S.; Pea, G.; Fengler, K.; Zuccolo, A.; Rafalski, A. (2005). "Gene duplication and exon shuffling by helitron-like transposons generate intraspecies diversity in maize". Nature Genetics. 37 (9): 997–1002. doi:10.1038/ng1615. PMID 16056225

5

Micropropagation and Plant Hormone

Micropropagation is a process of choosing the kind of plant material that needs to be propagated. The steps taken in this procedure are establishment, multiplication, pretransplant and transfer from culture. In order to completely understand micropropagation and plant hormone, it is necessary to understand the processes related to it. The following chapter elucidates the varied processes and mechanisms associated with this area of study.

Micropropagation

A rose plant that began as cells grown in a tissue culture

Micropropagation is the practice of rapidly multiplying stock plant material to produce a large number of progeny plants, using modern plant tissue culture methods.

Micropropagation is used to multiply noble plants such as those that have been genetically modified or bred through conventional plant breeding methods. It is also used to provide a sufficient number of plantlets for planting from a stock plant which does not produce seeds, or does not respond well to vegetative reproduction.

Cornell University botanist Frederick Campion Steward discovered and pioneered micropropagation and plant tissue culture in the late 1950s and early 1960s.

Stages of Micropropagation

Establishment

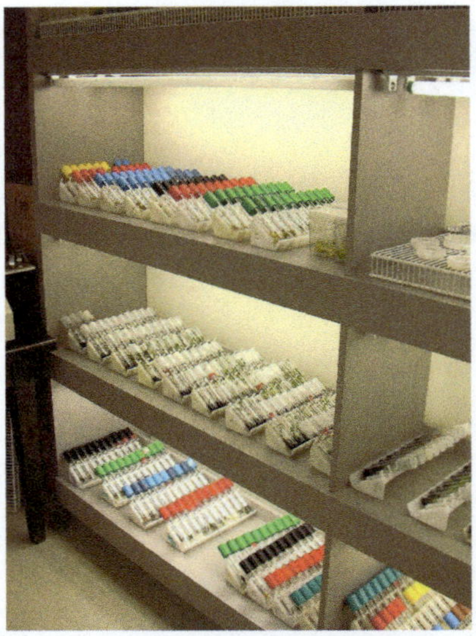

In vitro culture of plants in a controlled, sterile environment

Micropropagation begins with the selection of plant material to be propagated. The plant tissues are removed from an intact plant in a sterile condition. Clean stock materials that are free of viruses and fungi are important in the production of the healthiest plants. Once the plant material is chosen for culture, the collection of explant(s) begins and is dependent on the type of tissue to be used; including stem tips, anthers, petals, pollen and other plant tissues. The explant material is then surface sterilized, usually in multiple courses of bleach and alcohol washes, and finally rinsed in sterilized water. This small portion of plant tissue, sometimes only a single cell, is placed on a growth medium, typically containing sucrose as an energy source and one or more plant growth regulators (plant hormones). Usually the medium is thickened with agar to create a gel which supports the explant during growth. Some plants are easily grown on simple media, but others require more complicated media for successful growth; the plant tissue grows and differentiates into new tissues depending on the medium. For example, media containing cytokinin are used to create branched shoots from plant buds.

Multiplication

Multiplication is the taking of tissue samples produced during the first stage and increasing their number. Following the successful introduction and growth of plant tissue, the establishment stage is followed by multiplication. Through repeated cycles of this process, a single explant sample may be increased from one to hundreds and thousands of plants. Depending on the type of tissue grown, multiplication can involve different methods and media. If the plant material grown is callus tissue, it can be placed in a blender and cut into smaller pieces and recultured on the same type of culture medium to grow more callus tissue. If the tissue is grown as small plants called plantlets, hormones are often added that cause the plantlets to produce many small offshoots. After the formation of multiple shoots, these shoots are transferred to rooting medium with a high auxin\cytokinin ratio. After the development of roots, plantlets can be used for hardening.

Pretransplant

Banana plantlets transferred to soil (with vermicompost) from plant media. This process is done for acclimatization of plantlets to the soil as they were previously grown in plant media. After growing for some days the plantlets are transferred to the field.

This stage involves treating the plantlets/shoots produced to encourage root growth and "hardening." It is performed *in vitro*, or in a sterile "test tube" environment.

"Hardening" refers to the preparation of the plants for a natural growth environment. Until this stage, the plantlets have been grown in "ideal" conditions, designed to encourage rapid growth. Due to the controlled nature of their maturation, the plantlets often do not have fully functional dermal coverings. This causes them to be highly susceptible to disease and inefficient in their use of water and energy. In vitro conditions are high in humidity, and plants grown under these conditions often do not form a working cuticle and stomata that keep the plant from drying out. When taken out of culture, the plantlets need time to adjust to more natural environmental conditions. Hardening typically involves slowly weaning the plantlets from a high-humidity, low light, warm environment to what would be considered a normal growth environment for the species in question.

Transfer From Culture

Plant tissue cultures being grown at a USDA seed bank, the National Center for Genetic Resources Preservation.

In the final stage of plant micropropagation, the plantlets are removed from the plant media and transferred to soil or (more commonly) potting compost for continued growth by conventional methods.

This stage is often combined with the "pretransplant" stage.

Methods of Micro-propagation

Meristem Culture

In Meristem culture the Meristem and a few subtending leaf primordial are placed into a suitable growing media. An elongated rooted plantlet is produced after some weeks, and is transferred to the soil when it has attained a considerable height. A disease free plant can be produced by this method. Experimental result also suggest that this technique can be successfully utilized for rapid multiplication of various herbaceous Plants.

Callus Culture

A callus is mass of undifferentiated parenchymatous cells. When a living plant tissue is placed in an artificial growing medium with other conditions favorable, callus is formed. The growth of callus varies with the homogenous levels of auxin and Cyotkininn and can be manipulated by endogenous supply of these growth regulators in the culture medium. The callus growth and its organogenesis or embryogenesis can be referred into three different stages.

- Stage I: Rapid production of callus after placing the explants in culture medium
- Stage II: The callus is transferred to other medium containing growth regulators for the induction of adventitious organs.
- Stage III: The new plantlet is then exposed gradually to the environmental condition.

Suspension Culture

A cell suspension culture refers to cells and or groups of cells dispersed and growing in an aerated liquid culture medium (Street, 1997, Thorpe1981) is placed in a liquid medium and shaken vigorously and balanced dose of hormones. Suezawa et al. (1988) reported Cyotkininn induced adventitious buds in kiwi fruit in a suspension culture sub- culture for about a week.

Embryo Culture

In embryo culture, the embryo is excised and placed into a culture medium with proper nutrient in aseptic condition. To obtain a quick and optimum growth into plantlets, it is transferred to soil. It is particularly important for the production of interspecific and intergeneric hybrids and to overcome the embryo abortion.

Protoplast Culture

In protoplast culture, the plant cell can be isolated with the help of wall degrading enzymes and growth in a suitable culture medium in a controlled condition for regeneration of plantlets. Under suitable conditions the protoplast develops a cell wall followed by an increase in cell division and differentiation and grows into a new plant. The protoplast are first cultured in liquid medium at 25 to 28 C with a light intensity of 100 to 500 lux or in dark and after undergoing substantial cell division, they are transferred into solid medium congenial or morphogenesis in many horticultural crops response well to protoplast culture.

Advantages

Micropropagation has a number of advantages over traditional plant propagation techniques:

- The main advantage of micropropagation is the production of many plants that are clones of each other.
- Micropropagation can be used to produce disease-free plants.
- It can have an extraordinarily high fecundity rate, producing thousands of propagules while conventional techniques might only produce a fraction of this number.
- It is the only viable method of regenerating genetically modified cells or cells after protoplast fusion.
- It is useful in multiplying plants which produce seeds in uneconomical amounts, or when plants are sterile and do not produce viable seeds or when seed cannot be stored.
- Micropropagation often produces more robust plants, leading to accelerated growth compared to similar plants produced by conventional methods - like seeds or cuttings.
- Some plants with very small seeds, including most orchids, are most reliably grown from seed in sterile culture.
- A greater number of plants can be produced per square meter and the propagules can be stored longer and in a smaller area.

Disadvantages

Micropropagation is not always the perfect means of multiplying plants. Conditions that limits its use include:

- It is *very* expensive, and can have a labour cost of more than 70%.
- A monoculture is produced after micropropagation, leading to a lack of overall disease resilience, as all progeny plants may be vulnerable to the same infections.
- An infected plant sample can produce infected progeny. This is uncommon as the stock plants are carefully screened and vetted to prevent culturing plants infected with virus or fungus.
- Not all plants can be successfully tissue cultured, often because the proper medium for growth is not known or the plants produce secondary metabolic chemicals that stunt or kill the explant.
- Sometimes plants or cultivars do not come true to type after being tissue cultured. This is often dependent on the type of explant material utilized during the initiation phase or the result of the age of the cell or propagule line.
- Some plants are very difficult to disinfect of fungal organisms.

The major limitation in the use of micropropagation for many plants is the cost of production; for many plants the use of seeds, which are normally disease free and produced in good numbers, readily produce plants in good numbers at a lower cost. For this reason, many plant breeders do not utilize micropropagation because the cost is prohibitive. Other breeders use it to produce stock plants that are then used for seed multiplication.

Mechanisation of the process could reduce labour costs, but has proven difficult to achieve, despite active attempts to develop technological solutions.

Microspore

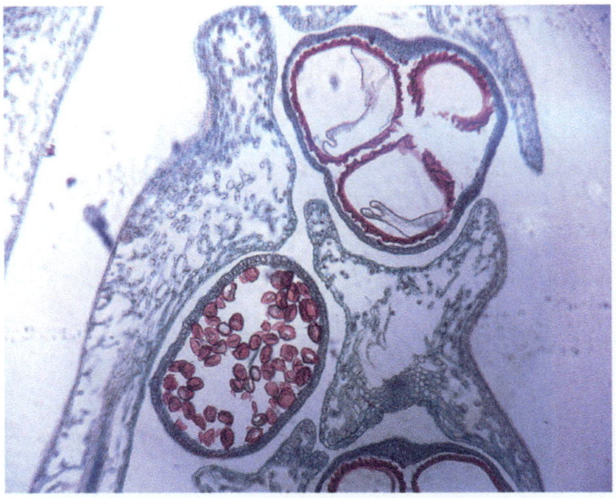

Microscopic photo of spores (in red) of *Selaginella*. The large three spores at the top are megaspores whereas the numerous smaller red spores at the bottom are microspores.

Microspores are land plant spores that develop into male gametophytes, whereas megaspores develop into female gametophytes. The male gametophyte gives rise to sperm cells, which are used for fertilization of an egg cell to form a zygote. Megaspores are structures that are part of the alternation of generations in many seedless vascular cryptogams, all gymnosperms and all angiosperms. Plants with heterosporous life cycles using microspores and megaspores arose independently in several plant groups during the Devonian period. Microspores are haploid, and are produced from diploid microsporocytes by meiosis.

Morphology

The microspore has three different wall layers. The outer layer is called the perispore, the next is the exospore, and the inner layer is the endospore. The perispore is the thickest of the three layers while the exospore and endospore are relatively equal in width.

Seedless Vascular Plants

In heterosporous seedless vascular plants, modified leaves called microsporophylls bear microsporangia containing many microsporocytes that undergo meiosis, each producing four microspores. Each microspore may develop into a male gametophyte consisting of a somewhat spherical antheridium within the microspore wall. Either 128 or 256 sperm cells with flagella are produced in each antheridium. The only heterosporous ferns are aquatic or semi-aquatic, including the genera

Marsilea,Regnellidium, *Pilularia*, *Salvinia*, and *Azolla*. Heterospory is also known in the lycopod genus *Selaginella* and in the quillwort genus *Isoëtes*.

Types of seedless vascular plants:

- Ferns
- Spikemosses
- Quillworts

Gymnosperms

In seed plants the microspores develop into pollen grains each containing a reduced, multicellular male gametophyte. The megaspores, in turn, develop into reduced female gametophytes that produce egg cells that, once fertilized, develop into seeds. Pollen cones or microstrobili usually develop toward the tips of the lower branches in clusters up to 50 or more. The microsporangia of gymnosperms develop in pairs toward the bases of the scales, which are therefore called microsporophylls. Each of the microsporocytes in the microsporangia undergoes meiosis, producing four haploid microspores. These develop into pollen grains, each consisting of four cells and a pair of external air sacs. The air sacs give the pollen grains added buoyancy that helps with wind dispersal.

Types of Gymnosperms:

- Conifers
  - Pines
- Ginkgos
- Cycads
- Gnetophytes

Angiosperms

As the anther of a flowering plant develops, four patches of tissue differentiate from the main mass of cells. These patches of tissue contain many diploid microsporocyte cells, each of which undergoes meiosis producing a quartet of microspores. Four chambers (pollen sacs) lined with nutritive tapetal cells are visible by the time the microspores are produced. After meiosis, the haploid microspores undergo several changes:

1. The microspore divides by mitosis producing two cells. The first of the cells (the generative cell) is small and is formed inside the second larger cell (the tube cell).
2. The members of each part of the microspores separate from each other.
3. A double-layered wall then develops around each microspore.

These steps occur at relatively the same time and when complete, the microspores have become pollen grains.

Embryogenesis

Although it is not the usual route of a microspore, this process is the most effective way of yielding haploid and double haploid plants through the use of male sex hormones. Under certain stressors such as heat or starvation, plants select for microspore embryogenesis. It was found that over 250 different species of angiosperms responded this way. In the anther, after a microspore undergoes microsporogenesis, it can deviate towards embryogenesis and become star-like microspores. The microspore can then go one of four ways: Become an embryogenic microspore, undergo callogenesis to organogenesis (haploid/double haploid plant), become a pollen-like structure or die.

Plant Tissue Culture

Plant tissue culture is a collection of techniques used to maintain or grow plant cells, tissues or organs under sterile conditions on a nutrient culture medium of known composition. Plant tissue culture is widely used to produce clones of a plant in a method known as micropropagation. Different techniques in plant tissue culture may offer certain advantages over traditional methods of propagation, including:

- The production of exact copies of plants that produce particularly good flowers, fruits, or have other desirable traits.
- To quickly produce mature plants.
- The production of multiples of plants in the absence of seeds or necessary pollinators to produce seeds.
- The regeneration of whole plants from plant cells that have been genetically modified.
- The production of plants in sterile containers that allows them to be moved with greatly reduced chances of transmitting diseases, pests, and pathogens.
- The production of plants from seeds that otherwise have very low chances of germinating and growing, i.e.: orchids and *Nepenthes*.
- To clear particular plants of viral and other infections and to quickly multiply these plants as 'cleaned stock' for horticulture and agriculture.

Plant tissue culture relies on the fact that many plant cells have the ability to regenerate a whole plant (totipotency). Single cells, plant cells without cell walls (protoplasts), pieces of leaves, stems or roots can often be used to generate a new plant on culture media given the required nutrients and plant hormones.

Techniques

Preparation of plant tissue for tissue culture is performed under aseptic conditions under HEPA filtered air provided by a laminar flow cabinet. Thereafter, the tissue is grown in sterile containers, such as petri dishes or flasks in a growth room with controlled temperature and light intensity.

Living plant materials from the environment are naturally contaminated on their surfaces (and sometimes interiors) with microorganisms, so their surfaces are sterilized in chemical solutions (usually alcohol and sodium or calcium hypochlorite) before suitable samples (known as explants) are taken. The sterile explants are then usually placed on the surface of a sterile solid culture medium, but are sometimes placed directly into a sterile liquid medium, particularly when cell suspension cultures are desired. Solid and liquid media are generally composed of inorganic salts plus a few organic nutrients, vitamins and plant hormones. Solid media are prepared from liquid media with the addition of a gelling agent, usually purified agar.

In vitro tissue culture of potato explants

The composition of the medium, particularly the plant hormones and the nitrogen source (nitrate versus ammonium salts or amino acids) have profound effects on the morphology of the tissues that grow from the initial explant. For example, an excess of auxin will often result in a proliferation of roots, while an excess of cytokinin may yield shoots. A balance of both auxin and cytokinin will often produce an unorganised growth of cells, or callus, but the morphology of the outgrowth will depend on the plant species as well as the medium composition. As cultures grow, pieces are typically sliced off and subcultured onto new media to allow for growth or to alter the morphology of the culture. The skill and experience of the tissue culturist are important in judging which pieces to culture and which to discard.

As shoots emerge from a culture, they may be sliced off and rooted with auxin to produce plantlets which, when mature, can be transferred to potting soil for further growth in the greenhouse as normal plants.

Regeneration Pathways

The specific differences in the regeneration potential of different organs and explants have various explanations. The significant factors include differences in the stage of the cells in the cell cycle, the availability of or ability to transport endogenous growth regulators, and the metabolic capabilities of the cells. The most commonly used tissue explants are the meristematic ends of the plants like the stem tip, axillary bud tip and root tip. These tissues have high rates of cell division and either concentrate or produce required growth regulating substances including auxins and cytokinins.

Shoot regeneration efficiency in tissue culture is usually a quantitative trait that often varies between plant species and within a plant species among subspecies, varieties, cultivars, or ecotypes. Therefore, tissue culture regeneration can become complicated especially when many regeneration procedures have to be developed for different genotypes within the same species.

The three common pathways of plant tissue culture regeneration are propagation from preexisting meristems (shoot culture or nodal culture), organogenesis and non-zygotic embryogenesis.

The propagation of shoots or nodal segments is usually performed in four stages for mass production of plantlets through in vitro vegetative multiplication but organogenesis is a common method of micropropagation that involves tissue regeneration of adventitious organs or axillary buds directly or indirectly from the explants. Non-zygotic embryogenesis is a noteworthy developmental pathway that is highly comparable to that of zygotic embryos and it is an important pathway for producing somaclonal variants, developing artificial seeds, and synthesizing metabolites. Due to the single cell origin of non-zygotic embryos, they are preferred in several regeneration systems for micropropagation, ploidy manipulation, gene transfer, and synthetic seed production. Nonetheless, tissue regeneration via organogenesis has also proved to be advantageous for studying regulatory mechanisms of plant development.

Choice of Explant

The tissue obtained from a plant to be cultured is called an explant.

Explants can be taken from many different parts of a plant, including portions of shoots, leaves, stems, flowers, roots, single undifferentiated cells and from many types of mature cells provided are they still contain living cytoplasm and nuclei and are able de-differentiate and resume cell division. This has given rise to the concept of totipotentency of plant cells. However this is not true for all cells or for all plants. In many species explants of various organs vary in their rates of growth and regeneration, while some do not grow at all. The choice of explant material also determines if the plantlets developed via tissue culture are haploid or diploid. Also the risk of microbial contamination is increased with inappropriate explants.

The first method involving the meristems and induction of multiple shoots is the preferred method for the micropropagation industry since the risks of somaclonal variation (genetic variation induced in tissue culture) are minimal when compared to the other two methods. Somatic embryogenesis is a method that has the potential to be several times higher in multiplication rates and is amenable to handling in liquid culture systems like bioreactors.

Some explants, like the root tip, are hard to isolate and are contaminated with soil microflora that become problematic during the tissue culture process. Certain soil microflora can form tight associations with the root systems, or even grow within the root. Soil particles bound to roots are difficult to remove without injury to the roots that then allows microbial attack. These associated microflora will generally overgrow the tissue culture medium before there is significant growth of plant tissue.

Some cultured tissues are slow in their growth. For them there would be two options: (i) Optimizing the culture medium; (ii) Culturing highly responsive tissues or varieties. Necrosis can spoil cultured tissues. Generally, plant varieties differ in susceptibility to tissue culture necrosis. Thus, by culturing highly responsive varieties (or tissues) it can be managed.

Aerial (above soil) explants are also rich in undesirable microflora. However, they are more easily removed from the explant by gentle rinsing, and the remainder usually can be killed by surface sterilization. Most of the surface microflora do not form tight associations with the plant tissue.

Such associations can usually be found by visual inspection as a mosaic, de-colorization or localized necrosis on the surface of the explant.

An alternative for obtaining uncontaminated explants is to take explants from seedlings which are aseptically grown from surface-sterilized seeds. The hard surface of the seed is less permeable to penetration of harsh surface sterilizing agents, such as hypochlorite, so the acceptable conditions of sterilization used for seeds can be much more stringent than for vegetative tissues.

Tissue cultured plants are clones. If the original mother plant used to produce the first explants is susceptible to a pathogen or environmental condition, the entire crop would be susceptible to the same problem. Conversely, any positive traits would remain within the line also.

Applications

Plant tissue culture is used widely in the plant sciences, forestry, and in horticulture. Applications include:

- The commercial production of plants used as potting, landscape, and florist subjects, which uses meristem and shoot culture to produce large numbers of identical individuals.
- To conserve rare or endangered plant species.
- A plant breeder may use tissue culture to screen cells rather than plants for advantageous characters, e.g. herbicide resistance/tolerance.
- Large-scale growth of plant cells in liquid culture in bioreactors for production of valuable compounds, like plant-derived secondary metabolites and recombinant proteins used as biopharmaceuticals.
- To cross distantly related species by protoplast fusion and regeneration of the novel hybrid.
- To rapidly study the molecular basis for physiological, biochemical, and reproductive mechanisms in plants, for example in vitro selection for stress tolerant plants.
- To cross-pollinate distantly related species and then tissue culture the resulting embryo which would otherwise normally die (Embryo Rescue).
- For chromosome doubling and induction of polyploidy, for example doubled haploids, tetraploids, and other forms of polyploids. This is usually achieved by application of antimitotic agents such as colchicine or oryzalin.
- As a tissue for transformation, followed by either short-term testing of genetic constructs or regeneration of transgenic plants.
- Certain techniques such as meristem tip culture can be used to produce clean plant material from virused stock, such as potatoes and many species of soft fruit.
- Production of identical sterile hybrid species can be obtained.

Laboratories

Although some growers and nurseries have their own labs for propagating plants by the technique of tissue culture, a number of independent laboratories provide custom propagation services. The Plant Tissue Culture Information Exchange lists many commercial tissue culture labs. Since plant tissue culture is a very labour-intensive process, this would be an important factor in determining which plants would be commercially viable to propagate in a laboratory.

Plant Hormone

Lack of the plant hormone auxin can cause abnormal growth (right)

Plant hormones (also known as phytohormones) are chemicals that regulate plant growth. In the United Kingdom, these are termed 'plant growth substances'.

Plant hormones are signal molecules produced within the plant, and occur in extremely low concentrations. Hormones regulate cellular processes in targeted cells locally and, moved to other locations, in other functional parts of the plant. Hormones also determine the formation of flowers, stems, leaves, the shedding of leaves, and the development and ripening of fruit. Plants, unlike animals, lack glands that produce and secrete hormones. Instead, each cell is capable of producing hormones. Plant hormones shape the plant, affecting seed growth, time of flowering, the sex of flowers, senescence of leaves, and fruits. They affect which tissues grow upward and which grow downward, leaf formation and stem growth, fruit development and ripening, plant longevity, and even plant death. Hormones are vital to plant growth, and, lacking them, plants would be mostly a mass of undifferentiated cells. So they are also known as growth factors or growth hormones. The term 'Phytohormone' was coined by Thimann in 1948.

Phytohormones are found not only in higher plants but in algae, showing similar functions, and in microorganisms, such as unicellular fungi and bacteria, but in these cases they play no hormonal or other immediate physiological role in the producing organism and can, thus, be regarded as secondary metabolites.

Characteristics

Phyllody on a purple coneflower (*Echinacea purpurea*), a plant development abnormality where leaf-like structures replace flower organs. It can be caused by hormonal imbalance, among other reasons.

The word hormone is derived from Greek, meaning *set in motion*. Plant hormones affect gene expression and transcription levels, cellular division, and growth. They are naturally produced within plants, though very similar chemicals are produced by fungi and bacteria that can also affect plant growth. A large number of related chemical compounds are synthesized by humans. They are used to regulate the growth of cultivated plants, weeds, and in vitro-grown plants and plant cells; these man-made compounds are called Plant Growth Regulators or PGRs for short. Early in the study of plant hormones, "phytohormone" was the commonly used term, but its use is less widely applied now.

Plant hormones are not nutrients, but chemicals that in small amounts promote and influence the growth, development, and differentiation of cells and tissues. The biosynthesis of plant hormones within plant tissues is often diffuse and not always localized. Plants lack glands to produce and store hormones, because, unlike animals — which have two circulatory systems (lymphatic and cardiovascular) powered by a heart that moves fluids around the body — plants use more passive means to move chemicals around their bodies. Plants utilize simple chemicals as hormones, which move more easily through their tissues. They are often produced and used on a local basis within the plant body. Plant cells produce hormones that affect even different regions of the cell producing the hormone.

Hormones are transported within the plant by utilizing four types of movements. For localized movement, cytoplasmic streaming within cells and slow diffusion of ions and molecules between cells are utilized. Vascular tissues are used to move hormones from one part of the plant to another; these include sieve tubes or phloem that move sugars from the leaves to the roots and flowers, and xylem that moves water and mineral solutes from the roots to the foliage.

Not all plant cells respond to hormones, but those cells that do are programmed to respond at specific points in their growth cycle. The greatest effects occur at specific stages during the cell's life, with diminished effects occurring before or after this period. Plants need hormones at very specific times during plant growth and at specific locations. They also need to disengage the effects that hormones have when they are no longer needed. The production of hormones occurs very often at sites of active growth within the meristems, before cells have fully differentiated. After production, they are sometimes moved to other parts of the plant, where they cause an immediate effect; or they can be stored in cells to be released later. Plants use different pathways to regulate internal hormone quantities and moderate their effects; they can regulate the amount of chemicals used to biosynthesize hormones. They can store them in cells, inactivate them, or cannibalise already-formed hormones by conjugating them with carbohydrates, amino acids, or peptides. Plants can also break down hormones chemically, effectively destroying them. Plant hormones frequently regulate the concentrations of other plant hormones. Plants also move hormones around the plant diluting their concentrations.

The concentration of hormones required for plant responses are very low (10^{-6} to 10^{-5} mol/L). Because of these low concentrations, it has been very difficult to study plant hormones, and only since the late 1970s have scientists been able to start piecing together their effects and relationships to plant physiology. Much of the early work on plant hormones involved studying plants that were genetically deficient in one or involved the use of tissue-cultured plants grown *in vitro* that were subjected to differing ratios of hormones, and the resultant growth compared. The earliest scientific observation and study dates to the 1880s; the determination and observation of plant hormones and their identification was spread-out over the next 70 years.

Classes of Plant Hormones

In general, it is accepted that there are five major classes of plant hormones, some of which are made up of many different chemicals that can vary in structure from one plant to the next. The chemicals are each grouped together into one of these classes based on their structural similarities and on their effects on plant physiology. Other plant hormones and growth regulators are not easily grouped into these classes; they exist naturally or are synthesized by humans or other organisms, including chemicals that inhibit plant growth or interrupt the physiological processes within plants. Each class has positive as well as inhibitory functions, and most often work in tandem with each other, with varying ratios of one or more interplaying to affect growth regulation.

The five major classes are:

Abscisic Acid Hormone

Abscisic acid (also called ABA) is one of the most important plant growth regulators. It was discovered and researched under two different names before its chemical properties were fully known, it was called *dormin* and *abscicin II*. Once it was determined that the two compounds are the same, it was named abscisic acid. The name "abscisic acid" was given because it was found in high concentrations in newly abscissed or freshly fallen leaves.

This class of PGR is composed of one chemical compound normally produced in the leaves of plants, originating from chloroplasts, especially when plants are under stress. In general, it acts as an inhibitory chemical compound that affects bud growth, and seed and bud dormancy. It mediates changes within the apical meristem, causing bud dormancy and the alteration of the last set of leaves into protective bud covers. Since it was found in freshly abscissed leaves, it was thought to play a role in the processes of natural leaf drop, but further research has disproven this. In plant species from temperate parts of the world, it plays a role in leaf and seed dormancy by inhibiting growth, but, as it is dissipated from seeds or buds, growth begins. In other plants, as ABA levels decrease, growth then commences as gibberellin levels increase. Without ABA, buds and seeds would start to grow during warm periods in winter and be killed when it froze again. Since ABA dissipates slowly from the tissues and its effects take time to be offset by other plant hormones, there is a delay in physiological pathways that provide some protection from premature growth. It accumulates within seeds during fruit maturation, preventing seed germination within the fruit, or seed germination before winter. Abscisic acid's effects are degraded within plant tissues during cold temperatures or by its removal by water washing in out of the tissues, releasing the seeds and buds from dormancy.

In plants under water stress, ABA plays a role in closing the stomata. Soon after plants are water-stressed and the roots are deficient in water, a signal moves up to the leaves, causing the formation of ABA precursors there, which then move to the roots. The roots then release ABA, which is translocated to the foliage through the vascular system and modulates the potassium and sodium uptake within the guard cells, which then lose turgidity, closing the stomata. ABA exists in all parts of the plant and its concentration within any tissue seems to mediate its effects and function as a hormone; its degradation, or more properly catabolism, within the plant affects metabolic reactions and cellular growth and production of other hormones. Plants start life as a seed with high ABA levels. Just before the seed germinates, ABA levels decrease; during germination and early growth of the seedling, ABA levels decrease even more. As plants begin to produce shoots with fully functional leaves, ABA levels begin to increase, slowing down cellular growth in more "mature" areas of the plant. Stress from water or predation affects ABA production and catabolism rates, mediating another cascade of effects that trigger specific responses from targeted cells. Scientists are still piecing together the complex interactions and effects of this and other phytohormones.

Auxins

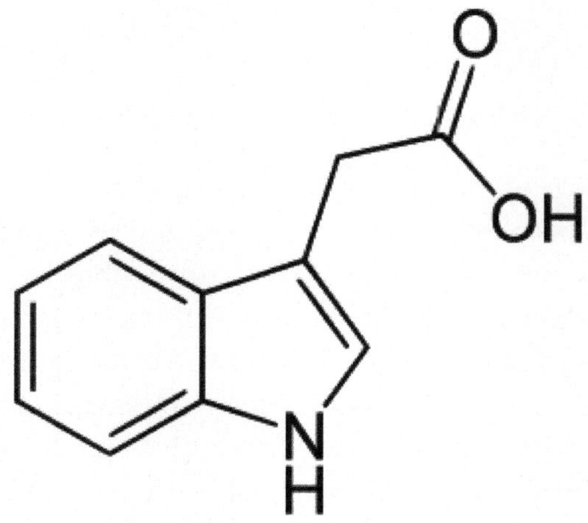

The auxin indole-3-acetic acid

Auxins are compounds that positively influence cell enlargement, bud formation and root initiation. They also promote the production of other hormones and in conjunction with cytokinins, they control the growth of stems, roots, and fruits, and convert stems into flowers. Auxins were the first class of growth regulators discovered. They affect cell elongation by altering cell wall plasticity. They stimulate cambium, a subtype of meristem cells, to divide and in stems cause secondary xylem to differentiate. Auxins act to inhibit the growth of buds lower down the stems (apical dominance), and also to promote lateral and adventitious root development and growth. Leaf abscission is initiated by the growing point of a plant ceasing to produce auxins. Auxins in seeds regulate specific protein synthesis, as they develop within the flower after pollination, causing the flower to develop a fruit to contain the developing seeds. Auxins are toxic to plants in large concentrations; they are most toxic to dicots and less so to monocots. Because of this property, synthetic auxin herbicides including 2,4-D and 2,4,5-T have been developed and used for weed control. Auxins, especially 1-Naphthaleneacetic acid (NAA) and Indole-3-butyric acid (IBA), are also commonly applied to stimulate root growth when taking cuttings of plants. The most common auxin found in plants is indole-3-acetic acid or IAA. The correlation of auxins and cytokinins in the plants is a constant (A/C = const.).

Cytokinins

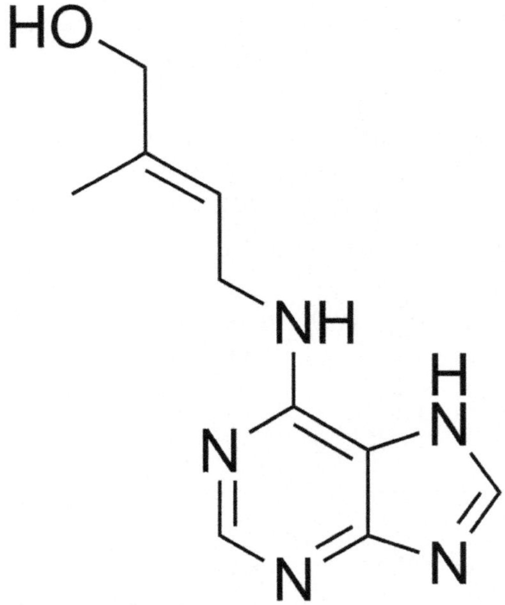

The cytokinin zeatin, the name is derived from *Zea* maize, in which it was first discovered in immature kernels.

Cytokinins or CKs are a group of chemicals that influence cell division and shoot formation. They were called kinins in the past when the first cytokinins were isolated from yeast cells. They also help delay senescence of tissues, are responsible for mediating auxin transport throughout the plant, and affect internodal length and leaf growth. They have a highly synergistic effect in concert with auxins, and the ratios of these two groups of plant hormones affect most major growth periods during a plant's lifetime. Cytokinins counter the apical dominance induced by auxins; they in conjunction with ethylene promote abscission of leaves, flower parts, and fruits. The correlation of auxins and cytokinins in the plants is a constant (A/C = const.).

Ethylene

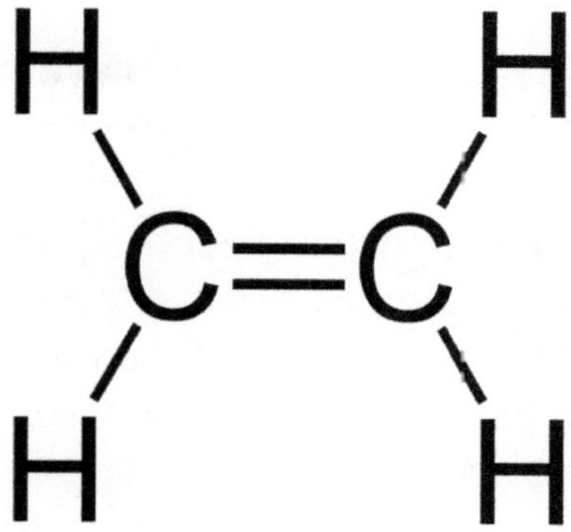

Ethylene

Ethylene is a gas that forms through the breakdown of methionine, which is in all cells. Ethylene has very limited solubility in water and does not accumulate within the cell but diffuses out of the cell and escapes out of the plant. Its effectiveness as a plant hormone is dependent on its rate of production versus its rate of escaping into the atmosphere. Ethylene is produced at a faster rate in rapidly growing and dividing cells, especially in darkness. New growth and newly germinated seedlings produce more ethylene than can escape the plant, which leads to elevated amounts of ethylene, inhibiting leaf expansion. As the new shoot is exposed to light, reactions by phytochrome in the plant's cells produce a signal for ethylene production to decrease, allowing leaf expansion. Ethylene affects cell growth and cell shape; when a growing shoot hits an obstacle while underground, ethylene production greatly increases, preventing cell elongation and causing the stem to swell. The resulting thicker stem can exert more pressure against the object impeding its path to the surface. If the shoot does not reach the surface and the ethylene stimulus becomes prolonged, it affects the stem's natural geotropic response, which is to grow upright, allowing it to grow around an object. Studies seem to indicate that ethylene affects stem diameter and height: When stems of trees are subjected to wind, causing lateral stress, greater ethylene production occurs, resulting in thicker, more sturdy tree trunks and branches. Ethylene affects fruit-ripening: Normally, when the seeds are mature, ethylene production increases and builds-up within the fruit, resulting in a climacteric event just before seed dispersal. The nuclear protein Ethylene Insensitive2 (EIN2) is regulated by ethylene production, and, in turn, regulates other hormones including ABA and stress hormones.

Gibberellins

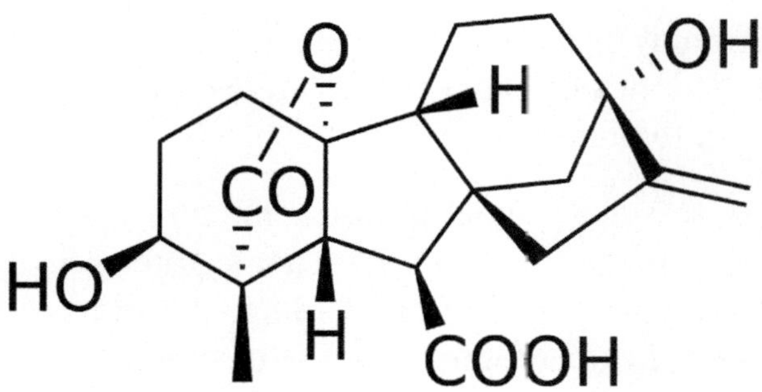

Gibberellin A1

Main function: initiate mobilization of storage materials in seeds during germination, cause elongation of stems, stimulate bolting in biennials stimulate pollen tube growth.

Gibberellins, or GAs, include a large range of chemicals that are produced naturally within plants and by fungi. They were first discovered when Japanese researchers, including Eiichi Kurosawa, noticed a chemical produced by a fungus called *Gibberella fujikuroi* that produced abnormal growth in rice plants. Gibberellins are important in seed germination, affecting enzyme production that mobilizes food production used for growth of new cells. This is done by modulating chromosomal transcription. In grain (rice, wheat, corn, etc.) seeds, a layer of cells called the aleurone layer wraps around the endosperm tissue. Absorption of water by the seed causes production of GA. The GA is transported to the aleurone layer, which responds by producing enzymes that break down stored food reserves within the endosperm, which are utilized by the growing seedling. GAs produce bolting of rosette-forming plants, increasing internodal length. They promote flowering, cellular division, and in seeds growth after germination. Gibberellins also reverse the inhibition of shoot growth and dormancy induced by ABA.

Other Known Hormones

Other identified plant growth regulators include:

- Brassinosteroids - are a class of polyhydroxysteroids, a group of plant growth regulators. Brassinosteroids have been recognized as a sixth class of plant hormones, which stimulate cell elongation and division, gravitropism, resistance to stress, and xylem differentiation. They inhibit root growth and leaf abscission. Brassinolide was the first identified brassinosteroid and was isolated from extracts of rapeseed (*Brassica napus*) pollen in 1979.
- Salicylic acid — activates genes in some plants that produce chemicals that aid in the defense against pathogenic invaders.
- Jasmonates — are produced from fatty acids and seem to promote the production of defense proteins that are used to fend off invading organisms. They are believed to also have a role in seed germination, and affect the storage of protein in seeds, and seem to affect root growth.
- Plant peptide hormones — encompasses all small secreted peptides that are involved in cell-to-cell signaling. These small peptide hormones play crucial roles in plant growth and development, including defense mechanisms, the control of cell division and expansion, and pollen self-incompatibility.
- Polyamines — are strongly basic molecules with low molecular weight that have been found in all organisms studied thus far. They are essential for plant growth and development and affect the process of mitosis and meiosis.
- Nitric oxide (NO) — serves as signal in hormonal and defense responses (e.g. stomatal closure, root development, germination, nitrogen fixation, cell death, stress response). NO can be produced by a yet undefined NO synthase, a special type of nitrite reductase, nitrate reductase, mitochondrial cytochrome c oxidase or non enzymatic processes and regulate plant cell organelle functions (e.g. ATP synthesis in chloroplasts and mitochondria).

- Strigolactones - implicated in the inhibition of shoot branching.
- Karrikins - not plant hormones because they are not made by plants, but are a group of plant growth regulators found in the smoke of burning plant material that have the ability to stimulate the germination of seeds
- Triacontanol - a fatty alcohol that acts as a growth stimulant, especially initiating new basal breaks in the rose family. It is found in alfalfa (lucerne), bee's wax, and some waxy leaf cuticles.

Potential Medical Applications

Plant stress hormones activate cellular responses, including cell death, to diverse stress situations in plants. Researchers have found that some plant stress hormones share the ability to adversely affect human cancer cells. For example, sodium salicylate has been found to suppress proliferation of lymphoblastic leukemia, prostate, breast, and melanoma human cancer cells. Jasmonic acid, a plant stress hormone that belongs to the jasmonate family, induced death in lymphoblastic leukemia cells. Methyl jasmonate has been found to induce cell death in a number of cancer cell lines.

Hormones and Plant Propagation

Synthetic plant hormones or PGRs are commonly used in a number of different techniques involving plant propagation from cuttings, grafting, micropropagation, and tissue culture.

The propagation of plants by cuttings of fully developed leaves, stems, or roots is performed by gardeners utilizing auxin as a rooting compound applied to the cut surface; the auxins are taken into the plant and promote root initiation. In grafting, auxin promotes callus tissue formation, which joins the surfaces of the graft together. In micropropagation, different PGRs are used to promote multiplication and then rooting of new plantlets. In the tissue-culturing of plant cells, PGRs are used to produce callus growth, multiplication, and rooting.

Seed Dormancy

Plant hormones affect seed germination and dormancy by acting on different parts of the seed.

Embryo dormancy is characterized by a high ABA:GA ratio, whereas the seed has a high ABA sensitivity and low GA sensitivity. In order to release the seed from this type of dormancy and initiate seed germination, an alteration in hormone biosynthesis and degradation toward a low ABA/GA ratio, along with a decrease in ABA sensitivity and an increase in GA sensitivity, must occur.

ABA controls embryo dormancy, and GA embryo germination. Seed coat dormancy involves the mechanical restriction of the seed coat. This, along with a low embryo growth potential, effectively produces seed dormancy. GA releases this dormancy by increasing the embryo growth potential, and/or weakening the seed coat so the radical of the seedling can break through the seed coat. Different types of seed coats can be made up of living or dead cells, and both types can be influenced by hormones; those composed of living cells are acted upon after seed formation, whereas the seed coats composed of dead cells can be influenced by hormones during the formation of the seed coat. ABA affects testa or seed coat growth characteristics, including thickness, and effects the GA-mediated embryo growth potential. These conditions and effects occur during the formation of the seed, often in response to environmental conditions. Hormones also mediate endosperm dormancy: Endosperm in most seeds is composed of living tissue that can actively respond to hormones generated by the embryo. The endosperm often acts as a barrier to seed germination, playing a part in seed coat dormancy or in the germination process. Living cells respond to and also affect the ABA:GA ratio, and mediate cellular sensitivity; GA thus increases the embryo growth potential and can promote endosperm weakening. GA also affects both ABA-independent and ABA-inhibiting processes within the endosperm.

Auxin

Native Auxins

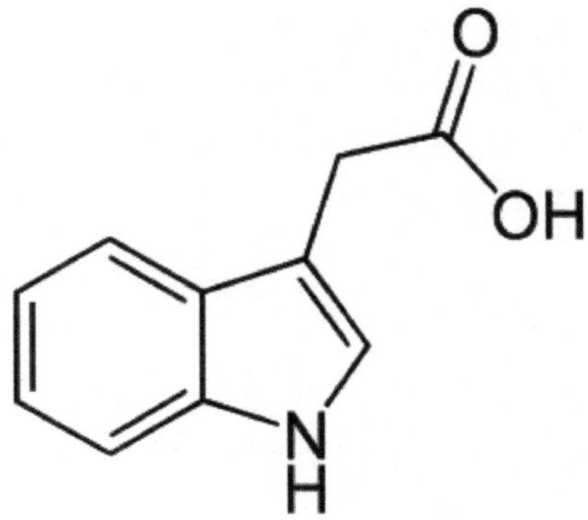

indole-3-acetic acid (IAA) is the most abundant and the basic auxin natively occurring and functioning in plants. It generates the majority of auxin effects in intact plants, and is the most potent native auxin.

There are four more endogenously synthesized auxins in plants.

All auxins are compounds with aromatic ring and a carboxylic acid group:

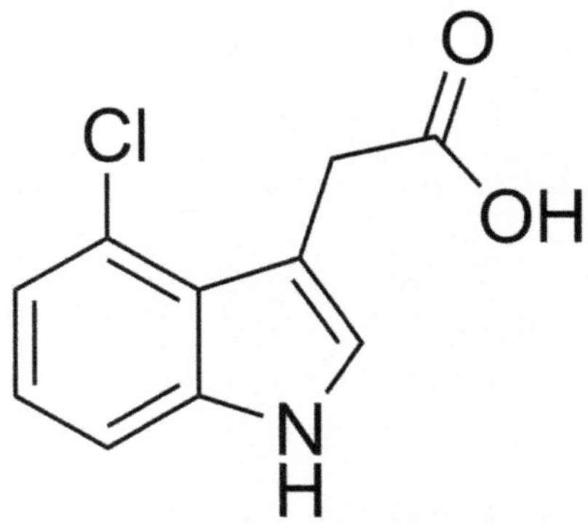

4-Chloroindole-3-acetic acid (4-CI-IAA)

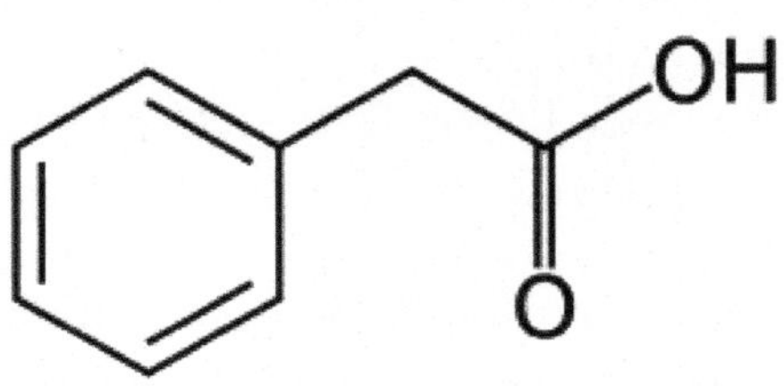

2-phenylacetic acid (PAA)

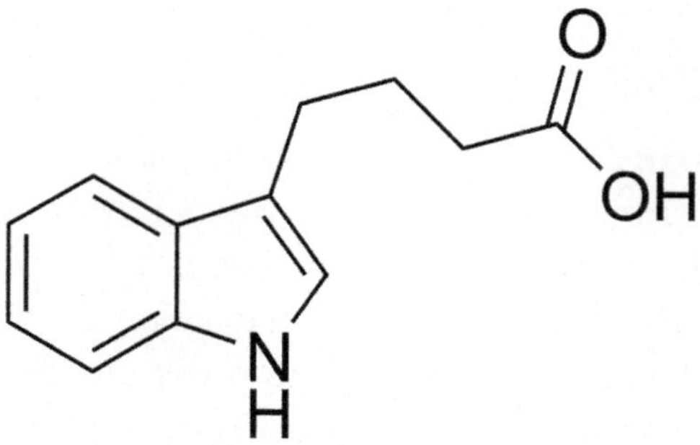

Indole-3-butyric acid (IBA)

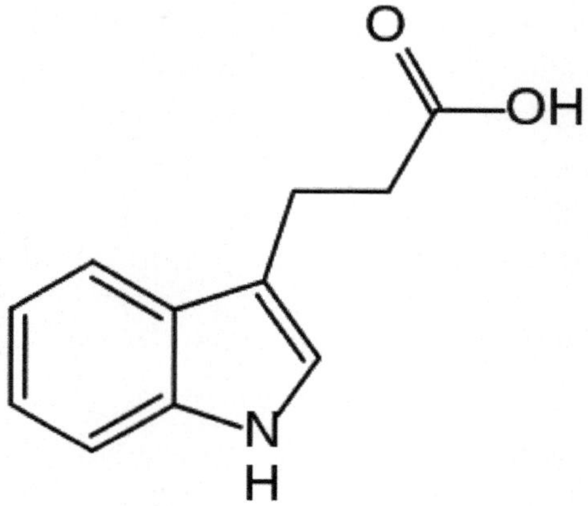

Indole-3-propionic acid (IPA)

Auxins are a class of plant hormones (or plant growth substances) with some morphogen-like characteristics. Auxins have a cardinal role in coordination of many growth and behavioral processes in the plant's life cycle and are essential for plant body development. Auxins and their role in plant growth were first described by the Dutch scientist Frits Warmolt Went. Kenneth V. Thimann was the first to isolate one of these phytohormones and determine its chemical structure as indole-3-acetic acid (IAA). Went and Thimann co-authored a book on plant hormones, *Phytohormones*, in 1937.

Overview

Auxins were the first of the major plant hormones to be discovered. They derive their name from the Greek word (*auxein* – "to grow/increase"). Auxin (namely IAA) is present in all parts of a plant, although in very different concentrations. The concentration in each position is crucial developmental information, so it is subject to tight regulation through both metabolism and transport. The result is the auxin creates "patterns" of auxin concentration maxima and minima in the plant body, which in turn guide further development of respective cells, and ultimately of the plant as a whole.

The (dynamic and environment responsive) pattern of auxin distribution within the plant is a key factor for plant growth, its reaction to its environment, and specifically for development of plant organs (such as leaves or flowers). It is achieved through very complex and well coordinated active transport of auxin molecules from cell to cell throughout the plant body — by the so-called polar auxin transport. Thus, a plant can (as a whole) react to external conditions and adjust to them, without requiring a nervous system. Auxins typically act in concert with, or in opposition to, other plant hormones. For example, the ratio of auxin to cytokinin in certain plant tissues determines initiation of root versus shoot buds.

On the molecular level, all auxins are compounds with an aromatic ring and a carboxylic acid group. The most important member of the auxin family is indole-3-acetic acid (IAA), which generates the majority of auxin effects in intact plants, and is the most potent native auxin. And as native auxin, its stability is controlled in many ways in plants, from synthesis, through possible conjugation to degradation of its molecules, always according to the requirements of the situation. However, molecules of IAA are chemically labile in aqueous solution, so it is not used commercially as a plant growth regulator.

- Five naturally occurring (endogenous) auxins in plants include indole-3-acetic acid, 4-chloroindole-3-acetic acid, phenylacetic acid, indole-3-butyric acid, and indole-3-propionic acid. However, most of the knowledge described so far in auxin biology, apply basically to IAA; the other three endogenous auxins seems to have rather marginal importance for intact plants in natural environments. Alongside endogenous auxins, scientists and manufacturers have developed many synthetic compounds with auxinic activity.
- Synthetic auxin analogs include 1-naphthaleneacetic acid, 2,4-dichlorophenoxyacetic acid (2,4-D), and many others.

Some synthetic auxins, such as 2,4-D and 2,4,5-trichlorophenoxyacetic acid (2,4,5-T), are used also as herbicides. Broad-leaf plants (dicots), such as dandelions, are much more susceptible to auxins than narrow-leaf plants (monocots) such as grasses and cereal crops, so these synthetic auxins are valuable as synthetic herbicides.

Discovery of Auxin

Charles Darwin

In 1881, Charles Darwin and his son Francis performed experiments on coleoptiles, the sheaths enclosing young leaves in germinating grass seedlings. The experiment exposed the coleoptile to light from a unidirectional source, and observed that they bend towards the light. By covering various parts of the coleoptiles with a light-impermeable opaque cap, the Darwins discovered that light is detected by the coleoptile tip, but that bending occurs in the hypocotyl. However the seedlings showed no signs of development towards light if the tip was covered with an opaque cap, or if the tip was removed. The Darwins concluded that the tip of the coleoptile was responsible for sensing light, and proposed that a messenger is transmitted in a downward direction from the tip of the coleoptile, causing it to bend.

Peter Boysen-Jensen

In 1913, Danish scientist Peter Boysen-Jensen demonstrated that the signal was not transfixed but mobile. He separated the tip from the remainder of the coleoptile by a cube of gelatine which prevented cellular contact, but allowed chemicals to pass through. The seedlings responded normally bending towards the light. However, when the tip was separated by an impermeable substance, there was no curvature of the stem.

Frits Went

In 1928, the Dutch botanist Frits Warmolt Went showed that a chemical messenger diffuses from coleoptile tips. Went's experiment identified how a growth promoting chemical causes a coleoptile to grow towards light. Went cut the tips of the coleoptiles and placed them in the dark, putting a few tips on agar blocks that he predicted would absorb the growth-promoting chemical. On control coleoptiles, he placed a block that lacked the chemical. On others, he placed blocks containing the chemical, either centred on top of the coleoptile to distribute the chemical evenly or offset to increase the concentration on one side.

When the growth-promoting chemical was distributed evenly the coleoptile grew straight. If the chemical was distributed unevenly, the coleoptile curved away from the side with the cube, as if growing towards light, even though it was grown in the dark. Went later proposed that the messenger substance is a growth-promoting hormone, which he named auxin, that becomes asymmetrically distributed in the bending region. Went concluded that auxin is at a higher concentration on the shaded side, promoting cell elongation, which results in a coleoptiles bending towards the light.

Hormonal Activity

Auxins help development at all levels in plants, from the cellular level, through organs, and ultimately to the whole plant.

Molecular Mechanisms

Auxin molecules present in cells may trigger responses directly through stimulation or inhibition of the expression of sets of certain genes or by means independent of gene expression. Auxin transcriptionally activates four different families of *early genes* (aka *primary response genes*), so-called because the components required for the activation are preexisting, leading to a rapid response. The families are glutathione S-transferases, auxin homeostasis proteins like GH3, SAUR genes of currently unknown function, and the Aux/IAA repressors.

Aux/IAA, ARF, TIR1, SCF Auxin Regulatory Pathways

The Aux/IAA repressors provide an example of one of the pathways leading to auxin induced changes of gene expression. This pathway involves the protein families TIR1 (transport inhibitor response1), ARF (auxin response factor), Aux/IAA transcriptional repressors, and the ubiquitin ligase complex that is a part of the ubiquitin-proteasome protein degradation pathway. ARF proteins have DNA binding domains and can bind promoter regions of genes and activate or repress gene expression. Aux/IAA proteins can bind ARF proteins sitting on gene promoters and prevent them from doing their job.

TIR1 proteins are F-box proteins that have three different domains giving them the ability to bind to three different ligands: an SCF^{TIR1} ubiquitin ligase complex (using the F-box domain), auxin (so TIR1 proteins are auxin receptors), and Aux/IAA proteins (via a degron domain). Upon binding of auxin, a TIR1 protein's degron domain has increased affinity for Aux/IAA repressor proteins, which when bound to TIR1 and its SCF complex undergo ubiquitination and subsequent degradation by a proteasome. The degradation of Aux/IAA proteins frees ARF proteins to activate or repress genes at whose promoters they are bound.

Within a plant, elaboration of the Aux/IAA repressor pathway takes place via diversification of the TIR1, ARF, and Aux/IAA protein families. Each family may contain many similar-acting proteins, differing in qualities such as degree of affinity for partner proteins, amount of activation or repression of target gene transcription, or domains of expression (e.g. different plant tissues might express different members of the family, or different environmental stresses might activate expression of different members). Such elaboration permits the plant to use auxin in a variety of ways depending on the needs of the tissue and plant.

Other Auxin Regulatory Pathways

Another protein, auxin-binding protein 1 (ABP1), is a putative receptor for a different signaling pathway, but its role is as yet unclear. Electrophysiological experiments with protoplasts and anti-ABP1 antibodies suggest ABP1 may have a function at the plasma membrane, and cells can possibly use ABP1 proteins to respond to auxin through means faster and independent of gene expression.

On a Cellular Level

On the cellular level, auxin is essential for cell growth, affecting both cell division and cellular expansion. Auxin concentration level, together with other local factors, contributes to cell differentiation and specification of the cell fate.

Depending on the specific tissue, auxin may promote axial elongation (as in shoots), lateral expansion (as in root swelling), or isodiametric expansion (as in fruit growth). In some cases (coleoptile growth), auxin-promoted cellular expansion occurs in the absence of cell division. In other cases, auxin-promoted cell division and cell expansion may be closely sequenced within the same tissue (root initiation, fruit growth). In a living plant, auxins and other plant hormones nearly always appear to interact to determine patterns of plant development.

Organ Patterns

Growth and division of plant cells together result in growth of tissue, and specific tissue growth contributes to the development of plant organs.

Growth of cells contributes to the plant's size, unevenly localized growth produces bending, turning and directionalization of organs- for example, stems turning toward light sources (phototropism), roots growing in response to gravity (gravitropism), and other tropisms originated because cells on one side grow faster than the cells on the other side of the organ. So, precise control of auxin distribution between different cells has paramount importance to the resulting form of plant growth and organization.

Auxin Transport and the Uneven Distribution of Auxin

To cause growth in the required domains, auxins must of necessity be active preferentially in them. Local auxin maxima can be formed by active biosynthesis in certain cells of tissues, for example via tryptophan-dependent pathways, but auxins are not synthesized in all cells (even if cells retain the potential ability to do so, only under specific conditions will auxin synthesis be activated in them). For that purpose, auxins have to be not only translocated toward those sites where they are needed, but also they must have an established mechanism to detect those sites. For that purpose, auxins have to be translocated toward those sites where they are needed. Translocation is driven throughout the plant body, primarily from peaks of shoots to peaks of roots (from up to down).

For long distances, relocation occurs via the stream of fluid in phloem vessels, but, for short-distance transport, a unique system of coordinated polar transport directly from cell to cell is exploited. This short-distance, active transport exhibits some morphogenetic properties.

This process, polar auxin transport, is directional, very strictly regulated, and based in uneven distribution of auxin efflux carriers on the plasma membrane, which send auxins in the proper direction. While PIN-FORMED (PIN) proteins are vital in transporting auxin in a polar manner, the family of AUXIN1/LIKE-AUX1 (AUX/LAX) genes encodes for non-polar auxin influx carriers.

The regulation of PIN protein localisation in a cell determines the direction of auxin transport from cell, and concentrated effort of many cells creates peaks of auxin, or auxin maxima (regions having cells with higher auxin – a maximum). Proper and timely auxin maxima within developing roots and shoots are necessary to organise the development of the organ. Surrounding auxin maxima are cells with low auxin troughs, or auxin minima. For example, in the *Arabidopsis* fruit, auxin minima have been shown to be important for its tissue development.

Organization of the Plant

As auxins contribute to organ shaping, they are also fundamentally required for proper development of the plant itself. Without hormonal regulation and organization, plants would be merely proliferating heaps of similar cells. Auxin employment begins in the embryo of the plant, where directional distribution of auxin ushers in subsequent growth and development of primary growth poles, then forms buds of future organs. Next, it helps to coordinate proper development of the arising organs, such as roots, cotyledons and leaves and mediates long distance signals between them, contributing so to the overall architecture of the plant. Throughout the plant's life, auxin helps the plant maintain the polarity of growth, and actually "recognize" where it has its branches (or any organ) connected.

An important principle of plant organization based upon auxin distribution is apical dominance, which means the auxin produced by the apical bud (or growing tip) diffuses (and is transported) downwards and inhibits the development of ulterior lateral bud growth, which would otherwise compete with the apical tip for light and nutrients. Removing the apical tip and its suppressively acting auxin allows the lower dormant lateral buds to develop, and the buds between the leaf stalk and stem produce new shoots which compete to become the lead growth. The process is actually quite complex, because auxin transported downwards from the lead shoot tip has to interact with several other plant hormones (such as strigolactones or cytokinins) in the process on various positions along the growth axis in plant body to achieve this phenomenon. This plant behavior is used in pruning by horticulturists.

Finally, the sum of auxin arriving from stems to roots influences the degree of root growth. If shoot tips are removed, the plant does not react just by outgrowth of lateral buds — which are supposed to replace to original lead. It also follows that smaller amount of auxin arriving to the roots results in slower growth of roots and the nutrients are subsequently in higher degree invested in the upper part of the plant, which hence starts to grow faster.

Effects

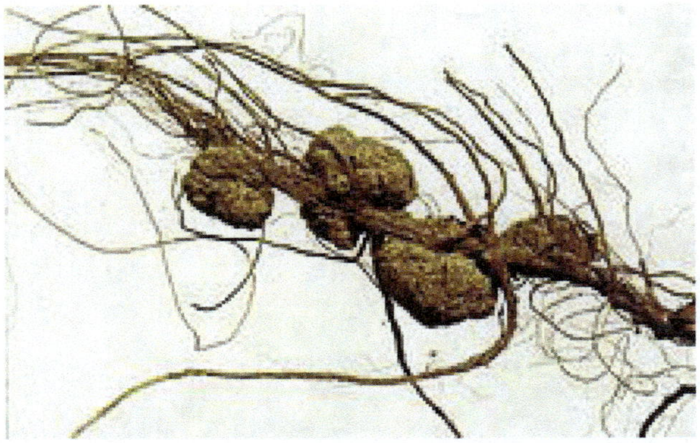

Crown galls are caused by *Agrobacterium tumefaciens* bacteria; they produce and excrete auxin and cytokinin, which interfere with normal cell division and cause tumors.

Auxin participates in phototropism, geotropism, hydrotropism and other developmental changes. The uneven distribution of auxin, due to environmental cues, such as unidirectional light or gravity force, results in uneven plant tissue growth, and generally, auxin governs the form and shape of plant body, direction and strength of growth of all organs, and their mutual interaction.

Auxin stimulates cell elongation by stimulating wall-loosening factors, such as elastins, to loosen cell walls. The effect is stronger if gibberellins are also present. Auxin also stimulates cell division if cytokinins are present. When auxin and cytokinin are applied to callus, rooting can be generated if the auxin concentration is higher than cytokinin concentration. Xylem tissues can be generated when the auxin concentration is equal to the cytokinins.

Auxin also induces sugar and mineral accumulation at the site of application.

Wound Response

Auxin induces the formation and organization of phloem and xylem. When the plant is wounded, the auxin may induce the cell differentiation and regeneration of the vascular tissues.

Root Growth and Development

Auxins promote root initiation. Auxin induces both growth of pre-existing roots and adventitious root formation, i.e., branching of the roots. As more native auxin is transported down the stem to the roots, the overall development of the roots is stimulated. If the source of auxin is removed, such as by trimming the tips of stems, the roots are less stimulated accordingly, and growth of stem is supported instead.

In horticulture, auxins, especially NAA and IBA, are commonly applied to stimulate root initiation when rooting cuttings of plants. However, high concentrations of auxin inhibit root elongation and instead enhance adventitious root formation. Removal of the root tip can lead to inhibition of secondary root formation.

Apical Dominance

Auxin induces shoot apical dominance; the axillary buds are inhibited by auxin, as a high concentration of auxin directly stimulates ethylene synthesis in axillary buds, causing inhibition of their growth and potentiation of apical dominance. When the apex of the plant is removed, the inhibitory effect is removed and the growth of lateral buds is enhanced. Auxin is sent to the part of the plant facing away from the light, where it promotes cell elongation, thus causing the plant to bend towards the light.

Fruit Growth and Development

Auxin is required for fruit growth and development and delays fruit senescence. When seeds are removed from strawberries, fruit growth is stopped; exogenous auxin stimulates the growth in fruits with seeds removed. For fruit with unfertilized seeds, exogenous auxin results in parthenocarpy ("virgin-fruit" growth).

Fruits form abnormal morphologies when auxin transport is disturbed. In *Arabidopsis* fruits, auxin controls the release of seeds from the fruit (pod). The valve margins are a specialised tissue in pods that regulates when pod will open (dehiscence). Auxin must be removed from the valve margin cells to allow the valve margins to form. This process requires modification of the auxin transporters (PIN proteins).

Flowering

Auxin plays also a minor role in the initiation of flowering and development of reproductive organs. In low concentrations, it can delay the senescence of flowers. A number of plant mutants have been described that affect flowering and have deficiencies in either auxin synthesis or transport. In maize, one example is bif2 barren inflorescence2.

Ethylene Biosynthesis

In low concentrations, auxin can inhibit ethylene formation and transport of precursor in plants; however, high concentrations can induce the synthesis of ethylene. Therefore, the high concentration can induce femaleness of flowers in some species.

Auxin inhibits abscission prior to formation of abscission layer, and thus inhibits senescence of leaves.

Synthetic Auxins

In the course of research on auxin biology, many compounds with noticeable auxin activity were synthesized. Many of them had been found to have economical potential for man-controlled growth and development of plants in agronomy. Synthetic auxins include the following compounds:

- Gallery of synthetic auxins

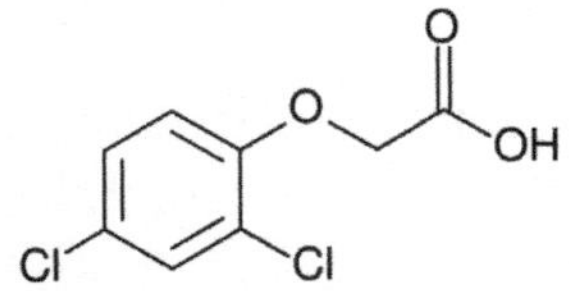	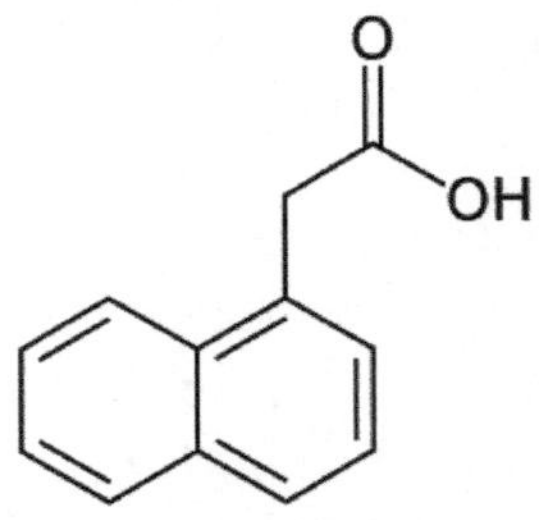	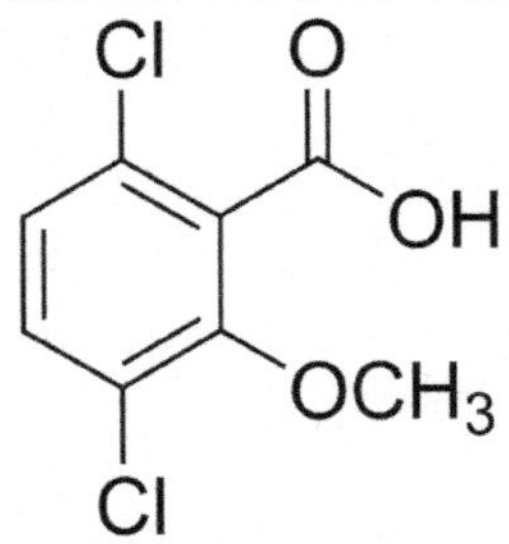
2,4-Dichlorophenoxyacetic acid (2,4-D); active herbicide and main auxin in laboratory use	α-Naphthalene acetic acid (α-NAA); often part of commercial rooting powders	2-Methoxy-3,6-dichlorobenzoic acid (dicamba); active herbicide

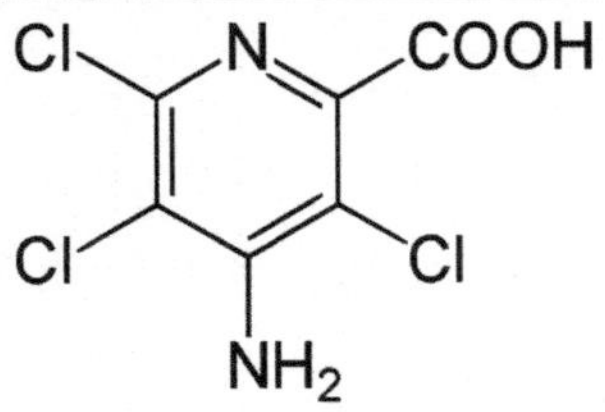	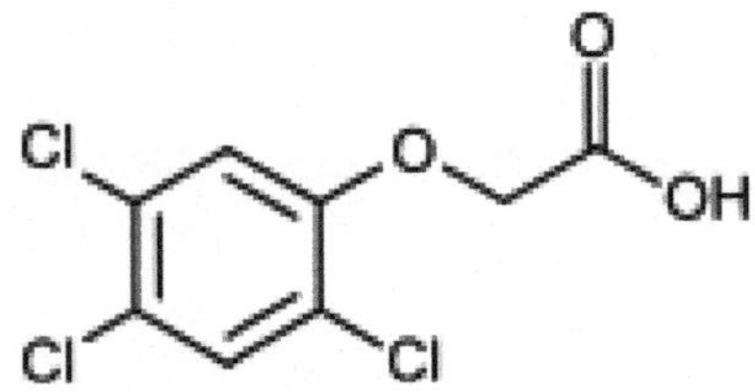
4-Amino-3,5,6-trichloropicolinic acid (tordon or picloram); active herbicide	2,4,5-Trichlorophenoxyacetic acid (2,4,5-T)

Auxins are toxic to plants in large concentrations; they are most toxic to dicots and less so to monocots. Because of this property, synthetic auxin herbicides, including 2,4-D and 2,4,5-T, have been developed and used for weed control.

However, some exogenously synthesized auxins, especially 1-naphthaleneacetic acid (NAA) and indole-3-butyric acid (IBA), are also commonly applied to stimulate root growth when taking cuttings of plants or for different agricultural purposes such as the prevention of fruit drop in orchards.

Used in high doses, auxin stimulates the production of ethylene. Excess ethylene (also native plant hormone) can inhibit elongation growth, cause leaves to fall (abscission), and even kill the plant. Some synthetic auxins, such as 2,4-D and 2,4,5-trichlorophenoxyacetic acid (2,4,5-T) were marketed also as herbicides. Dicots, such as dandelions, are much more susceptible to auxins than monocots, such as grasses and cereal crops. So, these synthetic auxins are valuable as synthetic herbicides. 2,4-D was the first widely used herbicide, and it is still so. 2,4-D was first commercialized by the Sherwin-Williams company, and saw use in the late 1940s. It is easy and inexpensive to manufacture.

Herbicide Manufacture

The defoliant Agent Orange, used extensively by British forces in the Malayan Emergency and American forces in the Vietnam War, was a mix of 2,4-D and 2,4,5-T. The compound 2,4-D is still in use and is thought to be safe, but 2,4,5-T was more or less banned by the U.S. Environmental Protection Agency in 1979. The dioxin TCDD is an unavoidable contaminant produced in the manufacture of 2,4,5-T. As a result of the integral dioxin contamination, 2,4,5-T has been implicated in leukemia, miscarriages, birth defects, liver damage, and other diseases.

Gibberellin

Gibberellins (GAs) are plant hormones that regulate growth and influence various developmental processes, including stem elongation, germination, dormancy, flowering, sex expression, enzyme induction, and leaf and fruit senescence.

Gibberellin was first recognized in 1926 by a Japanese scientist, Eiichi Kurosawa, studying *bakanae*, the "foolish seedling" disease in rice. It was first isolated in 1935 by Teijiro Yabuta and Sumuki, from fungal strain (*Gibberella fujikuroi*) provided by Kurosawa. Yabuta named the isolate as gibberellin.

Interest in gibberellins outside Japan began after World War II. In the United States, the first research was undertaken by a unit at Camp Detrick in Maryland, via studying seedlings of the bean *Vicia faba*. In the United Kingdom, work on isolating new types of gibberellin was undertaken at Imperial Chemical Industries. Interest in gibberellins spread around the world as the potential for its use on various commercially important plants became more obvious. For example, research that started at the University of California, Davis in the mid-1960s led to its commercial use on Thompson seedless table grapes throughout California by 1962. A known gibberellin biosynthesis inhibitor is paclobutrazol (PBZ), which in turn inhibits growth and induces early fruitset as well as seedset.

Chemistry

Effect of gibberellic acid on cannabis sprout

All known gibberellins are diterpenoid acids that are synthesized by the terpenoid pathway in plastids and then modified in the endoplasmic reticulum and cytosol until they reach their biologically-active form. All gibberellins are derived via the *ent*-gibberellane skeleton, but are synthesised via ent-kaurene. The gibberellins are named GA1 through GAn in order of discovery. Gibberellic acid, which was the first gibberellin to be structurally characterized, is GA3.

As of 2003, there were 126 GAs identified from plants, fungi, and bacteria.

Gibberellins are tetracyclic diterpene acids. There are two classes based on the presence of either 19 or 20 carbons. The 19-carbon gibberellins, such as gibberellic acid, have lost carbon 20 and, in place, possess a five-member lactone bridge that links carbons 4 and 10. The 19-carbon forms are, in general, the biologically active forms of gibberellins. Hydroxylation also has a great effect on the biological activity of the gibberellin. In general, the most biologically active compounds are dihydroxylated gibberellins, which possess hydroxyl groups on both carbon 3 and carbon 13. Gibberellic acid is a dihydroxylated gibberellin.

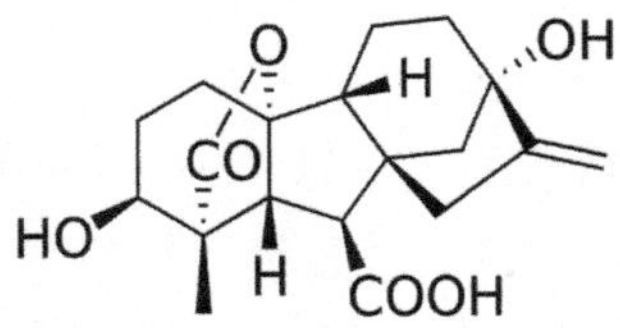

Gibberellin A1 (GA1)

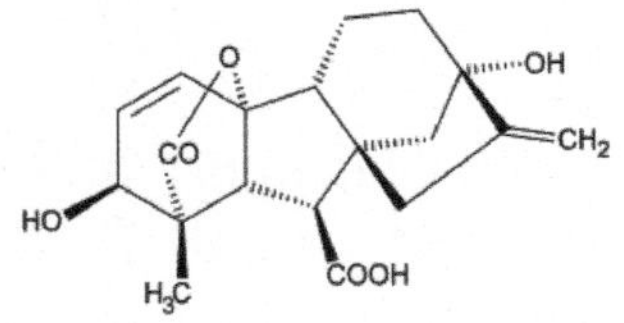

Gibberellic acid (GA3)

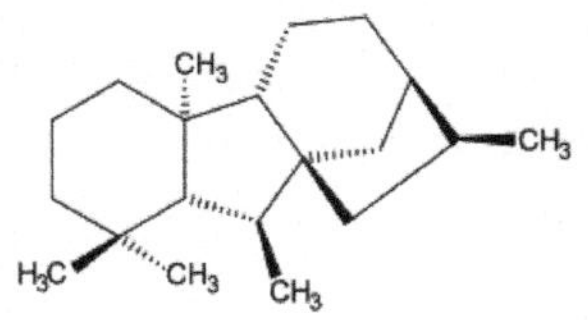

ent-Gibberellane

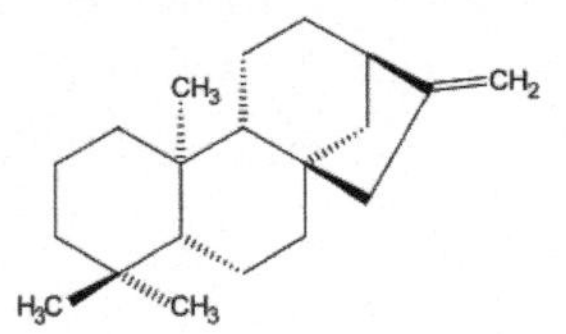

ent-Kaurene

Bioactive GAs

The bioactive GAs are GA1, GA3, GA4, and GA7. There are three common structural traits between these GAs: 1) a hydroxyl group on C-3β, 2) a carboxyl group on C-6, and 3) a lactone between C-4 and C-10. The 3β-hydroxyl group can be exchanged for other functional groups at C-2 and/or C-3 positions. GA5 and GA6 are examples of bioactive GAs that do not have a hydroxyl group on C-3β. The presence of GA1 in various plant species suggests that it is a common bioactive GA.

Biological Function

Gibberellins are involved in the natural process of breaking dormancy and other aspects of germination. Before the photosynthetic apparatus develops sufficiently in the early stages of germination, the stored energy reserves of starch nourish the seedling. Usually in germination, the breakdown of starch to glucose in the endosperm begins shortly after the seed is exposed to water. Gibberellins in the seed embryo are believed to signal starch hydrolysis through inducing the synthesis of the enzyme α-amylase in the aleurone cells. In the model for gibberellin-induced production of α-amylase, it is demonstrated that gibberellins (denoted by GA) produced in the scutellum diffuse to the aleurone cells, where they stimulate the secretion α-amylase. α-Amylase then hydrolyses starch, which is abundant in many seeds, into glucose that can be used in cellular respiration to produce energy for the seed embryo. Studies of this process have indicated gibberellins cause higher levels of transcription of the gene coding for the α-amylase enzyme, to stimulate the synthesis of α-amylase.

Gibberellins are produced in greater mass when the plant is exposed to cold temperatures. They stimulate cell elongation, breaking and budding, seedless fruits, and seed germination. They do the last by breaking the seed's dormancy and acting as a chemical messenger. Its hormone binds to a receptor, and Ca^{2+} activates the protein calmodulin, and the complex binds to DNA, producing an enzyme to stimulate growth in the embryo.

A major effect of gibberellins is the degradation of DELLA proteins, the absence of which then allows phytochrome interacting factors to bind to gene promoters and regulate gene expression. Gibberellins are thought to cause DELLAs to become polyubiquitinated and, thus, destroyed by the 26S proteasome pathway.

1. Shows a plant lacking gibberellins and has an internode length of "0" as well as it is a dwarf plant.
2. Shows your average plant with a moderate amount of gibberellins and an average internode length.
3.Shows a plant with a large amount of gibberellins and so has a much longer internode length because gibberellins promotes cell division in the stem.

Biosynthesis

GAs are usually produced from the methylerythritol phosphate (MEP) pathway in higher plants. In this pathway, bioactive GA is produced from trans-geranylgeranyl diphosphate (GGDP). In the MEP pathway, three classes of enzymes are used to yield GA from GGDP: 1) terpene synthases (TPSs), 2) cytochrome P450 monooxygenases (P450s), and 3) 2-oxoglutarate–dependent dioxygenases (2ODDs). There are 8 steps in the methylerythritol phosphate pathway: - 1) GGDP is converted to ent-copalyl diphosphate (ent-CPD) by ent-copalyl diphosphate synthase - 2) etn-CDP is converted to ent-kaurene by ent-kaurene synthase - 3) ent-kaurene is converted to ent-kaurenol by ent-kaurene oxidase (KO) - 4) ent-kaurenol is converted to ent-kaurenal by KO - 5) ent-kaurenal is converted to ent-kaurenoic acid by KO - 6) ent-kaurenoic acid is converted to ent-7a-hydroxykaurenoic acid by ent-kaurene acid oxidase (KAO) - 7) ent-7a-hydroxykaurenoic acid is converted to GA12-aldehyde by KAO - 8) GA12-aldehyde is converted to GA12 by KAO. GA12 is processed to the bioactive GA4 by oxidations on C-20 and C-3, which is accomplished by 2 soluble ODDs: GA 20-oxidase and GA 3-oxidase.

Sites of Biosynthesis

Most bioactive GAs are located in actively growing organs on plants. Both GA20ox and GA3ox genes (genes coding for GA 20-oxidase and GA 3-oxidase) and the SLENDER1 gene (a GA signal transduction gene) are found in growing organs on rice, which suggests bioactive GA synthesis occurs at their site of action in growing organs in plants. During flower development, the tapetum of anthers is believed to be a primary site of GA biosynthesis.

Differences Between Biosynthesis in Fungi and Lower Plants

Arabidopsis, a plant, and Gibberella fujikuroi, a fungus, possess different GA pathways and enzymes. P450s in fungi perform functions analogous to the functions of KAOs in plants. The function of CPS and KS in plants is performed by a single enzyme, CPS/KS, in fungi. In fungi, the GA biosynthesis genes are found on one chromosome, but in plants, they are found randomly on multiple chromosomes. Plants produce low amount of GA3, therefore the GA3 is produced for industrial purposes by microorganisms. Industrially the gibberellic acid can be produced by submerged fermentation, but this process presents low yield with high production costs and hence higher sale value, nevertheless other alternative process to reduce costs of the GA3 production is Solid-State Fermentation (SSF) that allows the use of agro-industrial residues.

Gibberellin Metabolism Genes

One or two genes encode the enzymes responsible for the first steps of GA biosynthesis in Arabidopsis and rice. The null alleles of the genes encoding CPS, KS, and KO result in GA-deficient Arabidopsis dwarves. Multigene families encode the 2ODDs that catalyze the formation of GA12 to bioactive GA4.

AtGA3ox1 and AtGA3ox2, two of the four genes that encode GA3ox in Arabidopsis, affect vegetative development. Environmental stimuli regulate AtGA3ox1 and AtGA3ox2 activity during seed germination. In Arabidopsis, GA20ox overexpression leads to an increase in GA concentration.

Deactivation

Several mechanisms for inactivating GAs have been identified. 2β-hydroxylation deactivates GA, and is catalyzed by GA2-oxidases (GA2oxs). Some GA2oxs use C19-GAs as substrates, and other GA2oxs use C20-GAs.

Cytochrome P450 mono-oxygenase, encoded by elongated uppermost internode (eui), converts GAs into 16α,17-epoxides. Rice eui mutants amass bioactive GAs at high levels, which suggests cytochrome P450 mono-oxygenase is a main enzyme responsible for deactivation GA in rice.

The Gamt1 and gamt2 genes encode enzymes that methylate the C-6 carboxyl group of GAs. In a gamt1 and gamt2 mutant, concentrations of GA is developing seeds is increased.

Regulation

Regulation by Other Hormones

The auxin indole-3-acetic acid (IAA) regulates concentration of GA1 in elongating internodes in peas. Removal of IAA by removal of the apical bud, the auxin source, reduces the concentration of GA1, and reintroduction of IAA reverses these effects to increase the concentration of GA1. This phenomenon has also been observed in tobacco plants. Auxin increases GA 3-oxidation and decreases GA 2-oxidation in barley. Auxin also regulates GA biosynthesis during fruit development in peas. These discoveries in different plant species suggest the auxin regulation of GA metabolism may be a universal mechanism.

Ethylene decreasing the concentration of bioactive GAs.

Regulation by Environmental Factors

Recent evidence suggests fluctuations in GA concentration influence light-regulated seed germination, photomorphogenesis during de-etiolation, and photoperiod regulation of stem elongation and flowering. Microarray analysis showed about one fourth cold-responsive genes are related to GA-regulated genes, which suggests GA influences response to cold temperatures. Plants reduce growth rate when exposed to stress. A relationship between GA levels and amount of stress experienced has been suggested in barley.

Role in Seed Development

Bioactive GAs and abcisic acid levels have an inverse relationship and regulate seed development and germination. Levels of FUS3, an Arabidopsis transcription factor, are upregulated by ABA and downregulated by GA, which suggests that there is a regulation loop that establishes the balance of GA and ABA.

Gibberellin Homeostasis

Feedback and feedforward regulation maintains the levels of bioactive GAs in plants. Levels of AtGA20ox1 and AtGA3ox1 expression are increased in a GA deficient environment, and decreased after the addition of bioactive GAs, Conversely, expression of AtGA2ox1 and AtGA2ox2, GA deactivation genes, is increased with addition of GA.

Impact on the "Green Revolution"

A chronic food shortage was feared during the rapid climb in world population in the 1960s. This was averted with the development of a high-yielding variety of rice. This variety of semi-dwarf rice is called IR8, and it has a short height because of a mutation in the sd1 gene. Sd1 encodes GA20ox, so a mutant sd1 is expected to exhibit a short height that is consistent with GA deficiency.

Receptors

GA receptors must have four traits: 1) reversibly bind GA, 2) GA saturability 3) high affinity for bioactive GA, and 4) reasonable ligand specificity for bioactive GAs.

Cereal Aleurone Cells

The first reported GA-binding protein (GBP) activity was discovered in wheat seed aleurone homogenates by using the GA-dependent induction of aleuron hydrolytic enzymes. Alpha-amylase induction in aleurone protoplasts is dependent on GA levels, but induction can still occur if GA does not pass through the plasma membrane. The existence of GA perception outside the cell and a plasma membrane GA receptor in aleurone cells can be concluded from these observations. Trimeric G proteins are involved in GA signaling, which supports the previous theory for plasma membrane GA receptors. Four different GBPs have been discovered in the plasma membrane of aleuron cells. Cereal aleurone cells could be unique in GA perception because they are some of the few plant cells that cannot synthesize bioactive GA.

GID1 Receptor

The rice GID1 gene encodes an unknown protein that only has a high affinity for bioactive GAs. In yeast, GID1 binds SLR1, GA signaling represor, depending on GA levels. GID1 is a soluble GA receptor mediating GA signaling. The GA-GID1 complex interacts with SLR1 to transduce the GA signal to SLR1. SLR1 increased the GA-binging activity of GID1 threefold. Arabidopsis has three GID1 homologs that bind GA and interact with DELLAs: AtGID1a, b, and c.

Gibberellin and DELLa Proteins

DELLA Proteins: Repressors of GA-dependent Processes

DELLA proteins act as intracellular repressors of GA responses, DELLAs inhibit seed germination, seed growth, and other GA-dependent pathways, but GA can reverse these effects.

GA-GID1-DELLA Complex

For GA to bind to the GID1 receptor, the C3-hydroxyl on GA must form a hydrogen bond to the tyrosine 31 residue of GID1, which induces a conformational change in GID1 to enclose GA. Once GA is trapped in the GID1 pocket, the lid of the pocket binds to DELLA to form the GA-GID1-DELLA complex.

GA Promotes Proteasome-dependent Degradation of DELLAs

In the absence of GA, many DELLAs are present to repress GA responses, but the formation of the GA-GID1-DELLA complex increases the degradation of DELLAs. The SCF complex is composed of SKP1, CULLIN, and F-BOX proteins. F-box proteins catalyze the formation of polyubiquitin on target proteins to be degraded by the 26S proteasome. The formation of the GA-GID1-DELLA complex is believed to cause a conformational change in DELLA, which enhances recognition of DELLAs by F-box proteins. Next, the SCF complex promotes ubiquitylation of DELLAs, which are then degraded by the 26S proteasome. Degradation of DELLAs allows GA regulated growth to resume. Thus, GA stimulates growth by activating the degradation of DELLAs.

Cytokinin

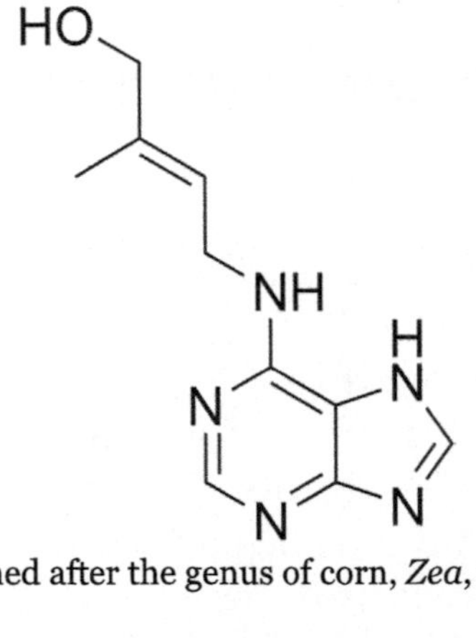

The cytokinin zeatin is named after the genus of corn, *Zea*, in which it was discovered.

Cytokinins (CK) are a class of plant growth substances (phytohormones) that promote cell division, or cytokinesis, in plant roots and shoots. They are involved primarily in cell growth and differentiation, but also affect apical dominance, axillary bud growth, and leaf senescence. Folke Skoog discovered their effects using coconut milk in the 1940s at the University of Wisconsin–Madison.

There are two types of cytokinins: adenine-type cytokinins represented by kinetin, zeatin, and 6-benzylaminopurine, and phenylurea-type cytokinins like diphenylurea and thidiazuron (TDZ). Most adenine-type cytokinins are synthesized in roots. Cambium and other actively dividing tissues also synthesize cytokinins. No phenylurea cytokinins have been found in plants. Cytokinins participate in local and long-distance signalling, with the same transport mechanism as purines and nucleosides. Typically, cytokinins are transported in the xylem.

Cytokinins act in concert with auxin, another plant growth hormone. The two are complementary, having generally opposite effects.

Mode of Action

The ratio of auxin to cytokinin plays an important role in the effect of cytokinin on plant growth. Cytokinin alone has no effect on parenchyma cells. When cultured with auxin but no cytokinin, they grow large but do not divide. When cytokinin is added, the cells expand and differentiate. When cytokinin and auxin are present in equal levels, the parenchyma cells form an undifferentiated callus. More cytokinin induces growth of shoot buds, while more auxin induces root formation.

Cytokinins are involved in many plant processes, including cell division and shoot and root morphogenesis. They are known to regulate axillary bud growth and apical dominance. The "direct inhibition hypothesis" posits that these effects result from the cytokinin to auxin ratio. This theory states that auxin from apical buds travels down shoots to inhibit axiliary bud growth. This promotes shoot growth, and restricts lateral branching. Cytokinin moves from the roots into the shoots, eventually signaling lateral bud growth. Simple experiments support this theory. When the apical bud is removed, the axillary buds are uninhibited, lateral growth increases, and plants become bushier. Applying auxin to the cut stem again inhibits lateral dominance.

While cytokinin action in vascular plants is described as pleiotropic, this class of plant hormones specifically induces the transition from apical growth to growth via a three-faced apical cell in moss protonema. This bud induction can be pinpointed to differentiation of a specific single cell, and thus is a very specific effect of cytokinin.

Cytokinins have been shown to slow aging of plant organs by preventing protein breakdown, activating protein synthesis, and assembling nutrients from nearby tissues. A study that regulated leaf senescence in tobacco leaves found that wild-type leaves yellowed while transgenic leaves remained mostly green. It was hypothesized that cytokinin may affect enzymes that regulate protein synthesis and degradation.

Cytokinin signaling in plants is mediated by a two-component phosphorelay. This pathway is initiated by cytokinin binding to a histidine kinase receptor in the endoplasmic reticulum membrane. This results in the autophosphorylation of the receptor, with the phosphate then being transferred to a phosphotransfer protein. The phosphotransfer proteins can then phosphorylate the type-B response regulators (RR) which are a family of transcriptions factors. The phosphorylated, and thus activated, type-B RRs regulate the transcription of numerous genes, including the type-A *RR*s. The type-A RRs negatively regulate the pathway.

Biosynthesis

Adenosine phosphate-isopentenyltransferase (IPT) catalyses the first reaction in the biosynthesis of isoprene cytokinins. It may use ATP, ADP, or AMP as substrates and may use dimethylallyl pyrophosphate (DMAPP) or hydroxymethylbutenyl pyrophosphate (HMBPP) as prenyl donors. This reaction is the rate-limiting step in cytokinin biosynthesis. DMADP and HMBDP used in cytokinin biosynthesis are produced by the methylerythritol phosphate pathway (MEP).

Cytokinins can also be produced by recycled tRNAs in plants and bacteria. tRNAs with anticodons that start with a uridine and carrying an already-prenylated adenosine adjacent to the anticodon release on degradation the adenosine as a cytokinin. The prenylation of these adenines is carried out by tRNA-isopentenyltransferase.

Auxin is known to regulate the biosynthesis of cytokinin.

Uses

Because cytokinin promotes plant cell division and growth, produce farmers use it to increase crops. One study found that applying cytokinin to cotton seedlings led to a 5–10% yield increase under drought conditions.

Cytokinins have recently been found to play a role in plant pathogenesis. For example, cytokinins have been described to induce resistance against *Pseudomonas syringae* in *Arabidopsis thaliana* and *Nicotiana tabacum*. Also in context of biological control of plant diseases cytokinins seem to have potential functions. Production of cytokinins by *Pseudomonas fluorescens* G20-18 has been identified as a key determinant to efficiently control the infection of *A. thaliana* with *P. syringae*.

Brassinosteroid

Brassinosteroids (BRs) are a class of polyhydroxysteroids that have been recognized as a sixth class of plant hormones. These were first explored nearly 40 years ago, when Mitchell et al. reported promotion in stem elongation and cell division by the treatment of organic extracts of rapeseed (*Brassica napus*) pollen. Brassinolide was the first isolated brassinosteroid in 1979, when pollen from *Brassica napus* was shown to promote stem elongation and cell divisions, and the biologically active molecule was isolated. The yield of brassinosteroids from 230 kg of *Brassica napus* pollen was only 10 mg. Since their discovery, over 70 BR compounds have been isolated from plants.

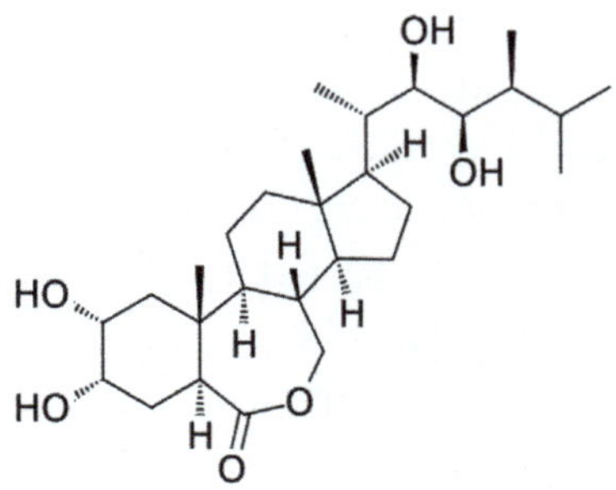

Brassinolide, the first brassinosteroid isolated and shown to have biological activity

Biosynthesis

The BR is biosynthesised from campesterol. The biosynthetic pathway was elucidated by Japanese researchers and later shown to be correct through the analysis of BR biosynthesis mutants in *Arabidopsis thaliana*, tomatoes, and peas. The sites for BR synthesis in plants have not been experimentally demonstrated. One well-supported hypothesis is that all tissues produce BRs, since BR biosynthetic and signal transduction genes are expressed in a wide range of plant organs, and short distance activity of the hormones also supports this. Experiments have shown that long distance transport is possible and that flow is in an acropetal direction, but it is not known if this movement is biologically relevant. Brassinosteroids are recognized at the cell membrane, although they are membrane-soluble.

Hormonal Activity

BRs have been shown to be involved in numerous plant processes:

- Promotion of cell expansion and cell elongation; works with auxin to do so.
- It has an unclear role in cell division and cell wall regeneration.
- Promotion of vascular differentiation; BR signal transduction has been studied during vascular differentiation.
- Is necessary for pollen elongation for pollen tube formation.
- Acceleration of senescence in dying tissue cultured cells; delayed senescence in BR mutants supports that this action may be biologically relevant.
- Can provide some protection to plants during chilling and drought stress.

Extract from the plant *Lychnis viscaria* contains a relatively high amount of Brassinosteroids. Lychnis viscaria increases the disease resistance of surrounding plants.

24-Epibrassinolide (EBL), a brassinosteroid isolated from *Aegle marmelos* Correa (Rutaceae), was further evaluated for the antigenotoxicity against maleic hydrazide (MH)-induced genotoxicity in *Allium cepa* chromosomal aberration assay. It was shown that the percentage of chromosomal aberrations induced by maleic hydrazide (0.01%) declined significantly with 24-epibrassinolide treatment.

BRs have been reported to counteract both abiotic and biotic stress in plants. Application of brassinosteroids to cucumbers was demonstrated to increase the metabolism and removal of pesticides, which could be beneficial for reducing the human ingestion of residual pesticides from non-organically grown vegetables. In all Type of brassinosteroids 28-homoBL is the most effective type of brassinosteroids. (sandeep kumar et al. 2010 Jour. of Indian bot society) Brassinosteroids increased tolerance to high temperature in Brassica juncea L. (Kumar S. 2010) The ability of 28-homobrassinolide to confer resistance to stress in Brassica juncea L. has also established (sandeep kumar). Application of 24-epiBL have any protective role on shoot, root length, soluble protein, proline content and peroxidases along with proline content PPO and IAA in seedlings of B. juncea L. under seasonal stress (Geetika Sirhindi).

BRs have also been reported to have a variety of effects when applied to rice seeds (Oryza sativa L.).Seeds treated with BRs were shown to reduce the growth inhibitory effect of salt stress. When the developed plants fresh weight was analyzed the treated seeds outperformed plants grown on saline and non-saline medium however when the dry weight was analyzed BR treated seeds only outperformed untreated plants that were grown on saline medium. When dealing with tomatoes (Lycopersicon esculentum) under salt stress the concentration of cholophyll a and cholophyll b were decreased and thus pigmentation was decreased as well. BR treated rice seeds considerably restored the pigment level in plants that were grown on saline medium when compared to non-treated plants under the same conditions.

Agricultural Uses

BR might reveal to have a prominent interest in the role of horticultural crops. Based on extensive research BR has the ability to improve the quantity and quality of horticultural crops and protect plants against many stresses that can be present in the local environment. With the many advances in technology dealing with the synthesis of more stable synthetic analogues and the genetic manipulation of cellular BR activity, using BR in the production of horticultural crops has become a more practical and hopeful strategy for improving crop yields and success.

BR could also help bridge the gap of the consumers' health concerns and the producers need for growth. A major benefit of using BR is that it does not interfere with the environment because they act in natural doses in a natural way. Since it is a "plant strengthening substance" and it is natural, BR application would be more favorable than pesticides and does not contribute to the co-evolution of pests.

In Germany, extract from the plant is allowed for use as a "plant strengthening substance."

Detection and Chemical Analysis

BRs can be detected by gas chromatography mass spectrometry and bioassays. There are some bioassays that can detect BRs in the plant such as the bean second internode elongation assay and the rice leaf lamina inclination test.

References

- Whippo, CW; Hangarter, RP (2006). "Phototropism: Bending towards enlightenment". The Plant cell. 18 (5): 1110–9. doi:10.1105/tpc.105.039669. PMC 1456868. PMID 16670442
- Tarakhovskaya, E. R.; Maslov, Yu; Shishova, M. F. (2007). "Phytohormones in algae". Russian Journal of Plant Physiology. 54 (2): 163–170. doi:10.1134/s1021443707020021
- Evert, Ray (2013). Biology of Plants. 41 Madison Avenue New York, NY 10010: Peter Marshall. ISBN 978-1-4292-1961-7
- Hohm, T; Preuten, T; Fankhauser, C (2013). "Phototropism: Translating light into directional growth". American journal of botany. 100 (1): 47–59. doi:10.3732/ajb.1200299. PMID 23152332
- Bidlack, James E.; Jansky, Shelley H. (2011). Stern's Introductory Plant Biology. New York, NY: McGraw-Hill. ISBN 0-07-304052-5
- Kermode AR (December 2005). "Role of Abscisic Acid in Seed Dormancy". J Plant Growth Regul. 24 (4): 319–344. doi:10.1007/s00344-005-0110-2

- MacMillan, J. (2001). "Occurrence of Gibberellins in Vascular Plants, Fungi, and Bacteria". Journal of plant growth regulation. 20 (4): 387–442. doi:10.1007/s003440010038. PMID 11986764
- Srivastava, L. M. (2002). Plant growth and development: hormones and environment. Academic Press. p. 140. ISBN 0-12-660570-X
- Sipes DL, Einset JW (August 1983). "Cytokinin stimulation of abscission in lemon pistil explants". J Plant Growth Regul. 2 (1–3): 73–80. doi:10.1007/BF02042235
- Hedden, G.; S. Thomas (2012). "Gibberellin biosynthesis and its regulation". The Biochemical Journal. 444 (11–25): 11–25. doi:10.1042/BJ20120245
- Öpik, Helgi; Rolfe, Stephen A.; Willis, Arthur John; Street, Herbert Edward (2005). The physiology of flowering plants (4th ed.). Cambridge University Press. p. 191. ISBN 978-0-521-66251-2
- Grennan, Aleel K (2006). "Gibberellin Metabolism Enzymes in Rice". Plant Physiology. 141 (2): 524–6. doi:10.1104/pp.104.900192. PMC 1475483. PMID 16760495
- Vettakkorumakankav, N. N.; D. Falk; P. Saxena; R. A. Fletcher (1999). "A Crucial Role for Gibberellins in Stress Protection of Plants". Plant and Cell Physiology. 40 (5): 542–548. doi:10.1093/oxfordjournals.pcp.a029575
- Weier, Thomas Elliot; Rost, Thomas L.; Weier, T. Elliot (1979). Botany: a brief introduction to plant biology. New York: Wiley. pp. 155–170. ISBN 0-471-02114-8
- Lindsey, Keith; Casson, Stuart; Chilley, Paul. (2002). "Peptides:new signalling molecules in plants". Trends in Plant Science. 7 (2): 78–83. doi:10.1016/S0960-9822(01)00435-3. PMID 11832279
- Achard, P.; P. Genschik (2009). "Releasing the brakes of plant growth: how GAs shutdown DELLA proteins". Journal of Experimental Botany. 60 (4): 1085–1092. doi:10.1093/jxb/ern301. PMID 19043067
- Osborne, Daphné J.; McManus, Michael T. (2005). Hormones, signals and target cells in plant development. Cambridge University Press. p. 158. ISBN 978-0-521-33076-3
- Simon, S; Petrášek, P (2011). "Why plants need more than one type of auxin". Plant Science. 180 (3): 454–460. doi:10.1016/j.plantsci.2010.12.007. PMID 21421392
- "Strigolactones and the control of plant development: Lessons from shoot branching". Plant Journal. 79: 607–622. doi:10.1111/tpj.12488
- Roszer T (2012) Nitric Oxide Synthesis in the Chloroplast. in: Roszer T. The Biology of Subcellular Nitric Oxide. Springer New York, London, Heidelberg. ISBN 978-94-007-2818-9
- Friml J (February 2003). "Auxin transport — shaping the plant". Current Opinion in Plant Biology. 6 (1): 7–12. doi:10.1016/S1369526602000031. PMID 12495745
- Vanneste, S; Friml, J (2012). "Plant signaling: Deconstructing auxin sensing". Nature Chemical Biology. 8 (5): 415–6. doi:10.1038/nchembio.943. PMID 22510662
- Mashiguchi, Kiyoshi (2011). "The main auxin biosynthesis pathway in Arabidopsis". PNAS. 108: 18512–7. doi:10.1073/pnas.1108434108. PMC 3215075. PMID 22025724

PERMISSIONS

All chapters in this book are published with permission under the Creative Commons Attribution Share Alike License or equivalent. Every chapter published in this book has been scrutinized by our experts. Their significance has been extensively debated. The topics covered herein carry significant information for a comprehensive understanding. They may even be implemented as practical applications or may be referred to as a beginning point for further studies.

We would like to thank the editorial team for lending their expertise to make the book truly unique. They have played a crucial role in the development of this book. Without their invaluable contributions this book wouldn't have been possible. They have made vital efforts to compile up to date information on the varied aspects of this subject to make this book a valuable addition to the collection of many professionals and students.

This book was conceptualized with the vision of imparting up-to-date and integrated information in this field. To ensure the same, a matchless editorial board was set up. Every individual on the board went through rigorous rounds of assessment to prove their worth. After which they invested a large part of their time researching and compiling the most relevant data for our readers.

The editorial board has been involved in producing this book since its inception. They have spent rigorous hours researching and exploring the diverse topics which have resulted in the successful publishing of this book. They have passed on their knowledge of decades through this book. To expedite this challenging task, the publisher supported the team at every step. A small team of assistant editors was also appointed to further simplify the editing procedure and attain best results for the readers.

Apart from the editorial board, the designing team has also invested a significant amount of their time in understanding the subject and creating the most relevant covers. They scrutinized every image to scout for the most suitable representation of the subject and create an appropriate cover for the book.

The publishing team has been an ardent support to the editorial, designing and production team. Their endless efforts to recruit the best for this project, has resulted in the accomplishment of this book. They are a veteran in the field of academics and their pool of knowledge is as vast as their experience in printing. Their expertise and guidance has proved useful at every step. Their uncompromising quality standards have made this book an exceptional effort. Their encouragement from time to time has been an inspiration for everyone.

The publisher and the editorial board hope that this book will prove to be a valuable piece of knowledge for students, practitioners and scholars across the globe.

Index